Pauli Lectures on Physics:
Thermodynamics and the Kinetic Theory of Gases

Pauli Lectures on Physics

1. Electrodynamics

2. Optics and the Theory of Electrons

3. Thermodynamics and the Kinetic Theory of Gases

4. Statistical Mechanics

5. Wave Mechanics

6. Selected Topics in Field Quantization

Pauli Lectures on Physics:
Volume 3. Thermodynamics and the Kinetic Theory of
Gases

Wolfgang Pauli

Edited by Charles P. Enz

Translated by H. R. Lewis and S. Margulies

Foreword by Victor F. Weisskopf

The MIT Press
Cambridge, Massachusetts, and London, England

ISBN 0 262 16048 x (hardcover)
Library of Congress catalog card number: 72–7803

Contents

Foreword

It is often said that scientific texts quickly become obsolete. Why are the Pauli lectures brought to the public today, when some of them were given as long as twenty years ago? The reason is simple: Pauli's way of presenting physics is never out of date. His famous article on the foundations of quantum mechanics appeared in 1933 in the German encyclopedia *Handbuch der Physik*. Twenty-five years later it reappeared practically unchanged in a new edition, whereas most other contributions to this encyclopedia had to be completely rewritten. The reason for this remarkable fact lies in Pauli's style, which is commensurate to the greatness of its subject in its clarity and impact. Style in scientific writing is a quality that today is on the point of vanishing. The pressure of fast publication is so great that people rush into print with hurriedly written papers and books that show little concern for careful formulation of ideas. Mathematical and instrumental techniques have become complicated and difficult; today most of the effort of writing and learning is devoted to the acquisition of these techniques instead of insight into important concepts. Essential ideas of physics are often lost in the dense forest of mathematical reasoning. This situation need not be so. Pauli's lectures show how physical ideas can be presented clearly and in good mathematical form, without being hidden in formalistic expertise.

Pauli was not an accomplished lecturer in the technical sense

of the word. It was often difficult to follow his courses. But
when the sequence of his thoughts and the structure of his
logic become apparent, the attentive follower is left with a new
and deeper knowledge of essential concepts and with a clearer
insight into the splendid architecture of reason, which is theoretical physics. The value of the lecture notes is not diminished
by the fact that they were written not by him but by some of
his collaborators. They bear the mark of the master in their
conceptual structure and their mathematical rigidity. Only
here and there does one miss words and comments of the master. Neither does one notice the passing of time in his lectures,
with the sole exception of the lectures on field quantization,
in which some concepts are formulated in a way that may
appear old-fashioned to some today. But even these lectures
should be of use to modern students because of their compact-
ness and their direct approach to the central problems.

May this volume serve as an example of how the concepts of
theoretical physics were conceived and taught by one of the
great men who created them.

Victor F. Weisskopf

Cambridge, Massachusetts

Preface

This is a conventional course on phenomenological thermodynamics. As Pauli says in the introduction, time does not
appear as variable in this framework, except for its direction.
Hence the subject is limited to the equilibrium theory (statics),
and irreversible thermodynamics is not discussed. But this conventional subject is treated by Pauli in the same inductive way
as the lectures on electrodynamics in this series, with the em-
phasis given to the historic development and the logical structure of the theory. It is for this reason that Pauli goes through
the labor of Carnot cycles and van't Hoff boxes. The axiomatic
formulation is only given as an illustration.

Occasionally Pauli's critical mind was stirred by his lecturing
even on an established subject like this. So in the last year of
his life, he sat back to think about a better formulation of
chemical reactions and, as mentioned in the appendix, a paper
grew out of this occupation with his course. This paper was
included in the second German edition, published in 1958, on
which this English translation is based. This edition was an
improved version of notes worked out by a student, E. Jucker,
and published in 1952.

Since Pauli taught this course at ETH in Zürich, equilibrium
thermodynamics has again become an active field of research.
Indeed, phase transitions are described today by scaling laws,
exhibiting the typical singularities of the thermodynamic quantities. This fascinating new field is beautifully described in a

very recent book by H. E. Stanley, *Introduction to Phase Transitions and Critical Phenomena* (Oxford University Press, New York, 1971). Furthermore, new techniques such as Raman scattering are applied to investigate thermodynamic systems, and new types of substances such as liquid crystals have become important. Therefore a solid basis of thermodynamics as given in these lectures is again more important today than at Pauli's time.

The last chapter, on kinetic theory of gases, is logically disconnected from the rest of the course. It actually belongs to the first chapter of the lectures on statistical mechanics in this series. The split in Pauli's lectures corresponds to the interruption between terms.

Although the editing of this course did not pose particular problems, the work of the translators deserves special acknowledgment.

Charles P. Enz

Geneva, 18 November 1971

Pauli Lectures on Physics:
Thermodynamics and the Kinetic Theory of Gases

Chapter 1. Basic Concepts and the First Law

Classical thermodynamics foregoes detailed pictures and, therefore, makes only general statements concerning the energetics of heat transfer. It limits itself to states of equilibrium and to very slowly occurring processes. No quantity with the dimension of time appears in thermodynamics; at the most, time enters via the concepts of "before" and "after." Therefore, in the case of rapidly occurring processes, only initial and final states are discussed. In thermodynamics no considerations concerning the nature of heat are made. This problem is first dealt with in the kinetic theory of gases.

1. THERMODYNAMIC VARIABLES

Definition: Thermodynamic variables are measurable macroscopic quantities which characterize a system.

Examples: Pressure p, volume V, surface tension σ, surface area A, stress tensor S_{ik}, strain tensor l_{ik}, magnetization M, magnetic field intensity H, concentration c, number of moles n.

2. CONCEPT OF TEMPERATURE

The temperature t of a system must first be defined. Experiment shows that if a system is closed, then heat is exchanged within the system until a stable thermal state is reached; this state is known as *thermodynamic equilibrium*.

1

Heat exchange through a heat-conducting connection makes thermodynamic equilibrium between two systems possible. Thus, we may say that two systems have the *same temperature* if they are in *thermodynamic equilibrium* (with one another).

A system in thermodynamic equilibrium possesses one less degree of freedom [A-1].[1] There exists a relation [A-2]

$$f(x_1, x_2, x_3, ..., x_n, y_1, y_2, y_3, ..., y_n) = \text{constant},$$

where x_i and y_i are the thermodynamic variables that characterize the system. In the simplest case of a homogeneous system this relation is

$$f(p, V) = \text{constant}.$$

If one variable in the above relation is held fixed, then the other variable is an arbitrary measure of temperature. For example, one such measure is the volume of a fixed amount of material at a fixed pressure. Since the variation of volume with temperature is different for different substances, temperature is not defined absolutely by volume at constant pressure. The exact *thermodynamic definition of the temperature scale* is not possible without the second law of thermodynamics. Nevertheless, our tentative definition always allows the determination of whether one temperature is larger or smaller than another: *Normal substances* expand with increasing temperature. However, there exist *anomalous substances* (for example, water between 0° and 4 °C), but these can be recognized as such by means of the following experiment. Two samples of a substance, A and B, which do not have the same temperature, are brought into thermal contact. During the subsequent heat exchange the samples experience volume changes ΔV_A and ΔV_B, respectively. For normal substances we always have $\Delta V_A \Delta V_B < 0$, but for anomalous subtances the initial conditions can be so

---

[1] Comments [A-1]–[A-13] appear in the appendix on pp. 130–132.

chosen that $\Delta V_A \Delta V_B > 0$. Steady-state heat conduction (refer to Section 3) offers a further possibility for deciding which of two temperatures is larger. By such means we obtain a monotonic temperature scale; that is, if $t_1 > t_2$ and we introduce a new scale through a monotonic transformation $t' = f(t)$, then $t_1' > t_2'$ also for the new temperatures [A-3].

3. QUANTITY OF HEAT

Experimentally we find that there exist equilibria between the phases of a substance:

$$\text{liquid} \rightleftarrows \text{gas},$$

$$\text{liquid} \rightleftarrows \text{solid},$$

$$\text{solid} \rightleftarrows \text{gas}.$$

In order to change the equilibrium between two phases, heat must be added or taken away; this heat is called the *heat of transformation* (e.g., heat of vaporization or heat of fusion). Temperature and pressure remain constant during the change. We can use the heat of transformation in defining *quantity of heat* (at constant temperature).

Definition: In order to change the phase of n grams of a substance, n times more heat is required than is needed to change the phase of 1 gram. (Heat of transformation is proportional to the amount of material.)

This definition is independent of the determination of the temperature scale, since the heat of transformation is added or subtracted at constant temperature.

For the comparison of quantities of heat at different temperatures, we use the process of steady-state heat conduction, in which two heat reservoirs with temperatures t_1 and t_2 $(t_1 > t_2)$ are connected by a heat conductor.

Definition: The quantity of heat Q_1 given up by reservoir 1 equals the quantity of heat Q_2 absorbed by reservoir 2.

The quantities Q_1 and Q_2 can also be heats of transformation, which then allows a direct comparison with the quantities of heat defined at constant temperature. A monotonic transformation of the temperature scale does not alter this definition either.

Historically, quantity of heat was defined by a mixing process. We mix two quantities of a substance, of masses m_1 and m_2 and temperatures t_1 and t_2 ($t_1 < t_2$), and obtain a mixture of mass $m_1 + m_2$ and temperature t_3. Equating the heat absorbed to the heat given up, we obtain

$$Q = cm_1(t_3 - t_1) = cm_2(t_2 - t_3) \,,$$

where c is the specific heat of the substance. From this it follows that

$$t_3 = \frac{m_1 t_1 + m_2 t_2}{m_1 + m_2} \,.$$

This formula is not invariant under the allowable transformations of the temperature scale. Likewise, the value of t_3 depends on whether p or V was held constant during the mixing. The above formula is correct only for sufficiently small temperature differences. If we set $t_3 = t_1 + dt$ and $t_2 = t_1 + dt_1$, we obtain

$$c(m_1 + m_2)\,dt = cm_2\,dt_1$$

or

$$dt = \frac{m_2}{m_1 + m_2}\,dt_1 \,.$$

An accurate definition of quantity of heat would then be

$$Q = m\int_0^t c(t)\,dt = m\int_0^{t'} c'(t')\,dt' \,,$$

since c, in general, depends on the temperature t. However, this definition of quantity of heat is very inconvenient. Since we already have defined quantity of heat, we can define c as the *specific heat* in the following way:

$$\delta Q = mc\,dt = mc'\,dt' \,.$$

Of course, the specific heat c depends on the definition of the temperature scale.

4. FIRST LAW OF THERMODYNAMICS

The first law gives the *connection between heat and other forms of energy.* We define *mechanical work* as work done by a system. It can always be reduced to the raising and lowering of weights.

Examples:

(*a*) If a homogeneous system is specified by the variables p and V (pressure and volume), then the work done by the system during a change of state is

$$\delta W = p \, dV .$$

(*b*) For a system specified by the surface tension σ and surface area A, the work done by the system during a change of state is

$$\delta W = -\sigma \, dA .$$

(*c*) Analogously, if a system is characterized by the magnetization $\boldsymbol{M}$ and magnetic field intensity $\boldsymbol{H}$, the work done by the system during a change of state is

$$\delta W = \boldsymbol{M} \cdot d\boldsymbol{H} .$$

Mechanical work is considered positive if done by the system when there is a positive change of state (that is, a positive change of the variables). The quantities p, σ, and $\boldsymbol{M}$ are called *intensive quantities* (y_i) because they occur as factors in the expressions for mechanical work. The quantities V, A, and $\boldsymbol{H}$ are called *extensive quantities* (x_i), because they occur as differentials. In general, mechanical work can be expressed as [A-2]

$$\delta W = \sum_{k=1}^{n-1} y_k(x_1, x_2, \ldots, x_n) \, dx_k$$

or

$$\delta W = \sum_{k=1}^{n} y_k(x_1, x_2, \ldots, x_n) \, dx_k , \qquad y_n = 0 .$$

Heat can be absorbed or released by a system without the occurrence of a mechanical displacement in reaction to a force. We consider δQ positive if heat is absorbed by the system.

The difference between heat and work cannot always be uniquely specified. It is assumed that there are cases, involving ideal processes, in which the two can be strictly distinguished from one another.

The First Law of Thermodynamics: If a system is taken from a state 1 to a state 2, then the sum of the heat added to the system and the work done on the system is independent of the path which leads from 1 to 2. That is,

$$J \sum_{(1 \to 2)} \delta Q - \sum_{(1 \to 2)} \delta W = f(1, 2) \, ,$$

where J is the mechanical equivalent of heat and depends on the system of units. (In cgs units, $J = 4.186 \times 10^7$ erg/cal.)[2] Henceforth, heat energy will always be measured in units of mechanical work so that J can be omitted. Thus, we may write

$$\sum_{(1 \to 2)} \delta Q - \sum_{(1 \to 2)} \delta W = f(1, 2) \, .$$

By choosing an arbitrary state o as the initial state, we can define the internal energy E_n of a state n as

$$\sum_{(o \to n)} \delta Q - \sum_{(o \to n)} \delta W = E_n \, .$$

Because of the arbitrary choice of the initial state, the internal energy is determined only to within a constant.

The first law can now be written in the following form:

$$E_2 - E_1 = \sum_{(1 \to 2)} \delta Q - \sum_{(1 \to 2)} \delta W \, .$$

Using the internal energy E_n, we can formulate the first

---

[2] Translator's Note: In the German edition, this value is incorrectly given as 4.136×10^7 erg/cal.

law more briefly by saying that E_n is a function of state (that is, it is path independent). For a cyclic process (identical initial and final states) we have

$$\sum_{(1\to1)} \delta Q - \sum_{(1\to1)} \delta W = 0 \,.$$

This equation states that during a cyclic process heat can only be transformed into work or vice versa. This is equivalent to the assertion that no cycling machine exists which produces heat energy or work from nothing, and it is a statement of the impossibility of constructing a perpetual motion machine of the first kind.

5. THERMODYNAMIC CHANGES OF STATE

A distinction is made between quasi-static (slowly occurring) and rapidly occurring changes of state on the one hand and between isothermal, isoenergetic, and adiabatic changes of state on the other.

a. Quasi-static

Not only is the result of the change of state reversible, but so also is each individual step.

Example: Gas in volume V_1 is infinitely slowly compressed to volume V_0. The gas and container are in an

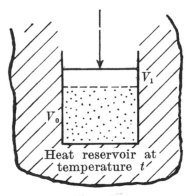

Figure 5.1

infinitely large heat reservoir, so that the temperature remains constant (isothermal change of state). There is no
change other than that work is done on the system. This
process can occur equally well in reverse, in that the gas
is expanded infinitely slowly from V_0 to V_1.

b. Rapidly occurring

These changes of state are controllable only if the system
is closed with respect to its surroundings.

Example: A gas contained in volume V_0 is allowed to
flow into an evacuated volume V_1, so that both V_0 and V_1

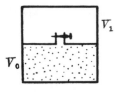

Figure 5.2

are filled with gas. This process occurs very rapidly and
is not reversible.

1. Isothermal: During the change of state, the tempera-
    ture t is constant ($dt = 0$).

2. Isoenergetic: During changes of state, the internal en-
    ergy E stays constant; that is, changes of state of
    completely closed systems are isoenergetic ($dE = 0$).

3. Adiabatic: During the change of state, no addition or
    removal of heat takes place; that is, the system is
    isolated by adiabatic walls (walls which do not con-
    duct heat) ($\delta Q = 0$).

It should be noted that the equations $dE = \delta Q - \delta W$ and
$\delta W = \sum_k y_k dx_k$ are well defined only for slow changes of state.

6. MATHEMATICAL FORMULATION OF THE FIRST LAW

In order that the expression $\sum_{k} y_k(x_1, x_2, ..., x_n)\, \mathrm{d}x_k$ be the exact differential $\mathrm{d}f$ of a function $f(x_1, x_2, ..., x_n)$, it is necessary that

$$\frac{\partial y_k}{\partial x_i} = \frac{\partial y_i}{\partial x_k} \qquad (i, k = 1, 2, ..., n).$$

These conditions follow from the following relations:

$$\frac{\partial f}{\partial x_k} = y_k \qquad \text{and} \qquad \frac{\partial y_k}{\partial x_i} = \frac{\partial^2 f}{\partial x_i \partial x_k} = \frac{\partial^2 f}{\partial x_k \partial x_i} = \frac{\partial y_i}{\partial x_k}.$$

On the other hand, Stokes' theorem states that

$$\int \sum_{i>k} \left(\frac{\partial y_i}{\partial x_k} - \frac{\partial y_k}{\partial x_i} \right) \mathrm{d}x_i \times \mathrm{d}x_k = \oint \sum_{k} y_k\, \mathrm{d}x_k.$$

If the conditions required above are fulfilled, then

$$\oint \sum_{k} y_k\, \mathrm{d}x_k = 0.$$

This means that the integral $\int_{1}^{2} \sum_{k} y_k \mathrm{d}x_k$ is path independent. Therefore, there exists a function $f(x_1, x_2, ..., x_n)$ which is determined to within a constant, and whose exact differential is the expression $\mathrm{d}f = \sum_{k} y_k \mathrm{d}x_k$. The required conditions are also sufficient.

Pfaff has shown that a differential form $\sum_{k} y_k\, \mathrm{d}x_k$ can always be transformed to the normal form $\sum_{\nu=1}^{m} X_{2\nu} \mathrm{d}X_{2\nu-1}$ $+ k\mathrm{d}X_{2m+1}$, where $2m \leqslant n$ (n even) or $2m+1 \leqslant n$ (n odd). If the number of variables n is even, then the term $\mathrm{d}X_{2m+1}$ is absent ($k = 0$). If $2m < n$ (n even) or if $2m+1 < n$ (n odd), then the system is degenerate. If we have a differential form $\delta h = y_1 \mathrm{d}x_1 + y_2 \mathrm{d}x_2$ with only two variables and

(*a*) if δh is an exact differential, then the normal form is

$$\delta h = \mathrm{d}X;$$

(b) if δh is not an exact differential, then we obtain

$$\delta h = y_1 \, dx_1 + y_2 \, dx_2 = X_2 \, dX_1 \quad \text{or} \quad \frac{\delta h}{X_2} = \frac{y_1 \, dx_1 + y_2 \, dx_2}{X_2} = dX_1 \, .$$

Thus, $\delta h / X_2 = dX_1$ is an exact differential, which means that $1/X_2$ is an integrating factor. Therefore, for a differential form with only two variables, there always exists an integrating factor.

In general, δQ *and δW are not exact differentials.* On the other hand, $\delta Q - \delta W = dE$ (differential of the internal energy) is, according to the first law, an exact differential. This is the same as saying that E_n is a function of state. As a consequence of the first law, all influences on a thermodynamic system can be reduced to the raising and lowering of weights and to the addition and removal of heat.

7. APPLICATIONS OF THE FIRST LAW

a. Definition of an ideal gas

An ideal gas is defined by the following three properties:

1   The internal energy $E(V, t)$ is independent of the volume V; that is,

$$E(V, t) = E(t) \quad \text{or} \quad \left(\frac{\partial E}{\partial V}\right)_t = 0 \, .$$

2   The isotherms are

$pV = \text{constant} \quad \text{(Law of Boyle and Mariotte)} \, .$ [7.1]

3   With a suitable temperature scale,

$$V_t = V_0(1 + \alpha t), \qquad\qquad [7.2]$$

where α is the same for all ideal gases. Therefore, with the help of the ideal gases, a new temperature scale can be introduced. It is, however, determined only to within a linear transformation. In the following we shall assume that such a gas scale has been introduced, for example, the Celsius scale ($\alpha = 1/273°$). Combining 2 and 3 we obtain [A-4]:

3'. $pV = p_0 V_0 f(t)$, where $f(t)$ is the same function for all gases: $f(t) = 1 + \alpha t$. If we set $1/\alpha + t = T$ (called the "absolute temperature"), then [7.2] simplifies to $V_t = V_0 \alpha T$ and we obtain

$$p_t V_t = p_0 V_0 \alpha T . \qquad [7.3]$$

It is useful to introduce the following quantities:

density: $\qquad \varrho = \dfrac{\text{mass}}{\text{volume}}$,

mole number: $n = \dfrac{\text{mass}}{\text{mass of one mole}}$,

molal volume: $v = \dfrac{\text{mass of one mole}}{\text{density}}$.

Since $p_0 V_0 \alpha$ is independent of temperature, pressure, and volume and only depends on the quantity of material, we may set $p_0 V_0 \alpha = Rn$; and since $V_t = nv$, we obtain

$$p_t V_t = p_t nv = p_0 V_0 \alpha T = RnT , \qquad [7.4]$$

or

$$pv = RT \qquad \text{(per mole)} . \qquad [7.5]$$

R is independent of the nature of the ideal gas and, in cgs units, $R = 8.31 \times 10^7$ erg/degree. The above formula is equivalent to Avogadro's law which states that the molal volumes of all gases at the same pressure and temperature are equal. Historically, this law led to the differentiation between atoms and molecules.

We cannot say how much a gas must be rarefied for it to be ideal. However, we can determine whether the gas is ideal with Joule's free-expansion experiment. The gas is allowed to stream into a vacuum and the temperature is measured before and after the experiment. The entire system must be completely isolated against energy transfer with the surroundings: $E_1 = E_0$. If $T_1 = T_0$, then it must also be that $(\partial E/\partial V)_T = 0$; that is, we have an ideal gas.

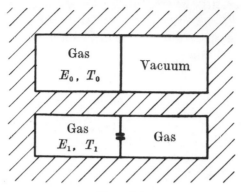

Figure 7.1

b. Specific heats

1. *Arbitrary homogeneous substances.* According to Section 3, the specific heat is calculated from [A-5]

$$c = \frac{\delta Q}{\mathrm{d}t}.$$

Depending on whether p or V is held constant, we distinguish between c_p and c_v, that is, the specific heat at constant pressure and that at constant volume, respectively. From the first law it follows that

$$c_V = \frac{(\delta Q)_V}{\mathrm{d}T} = \left(\frac{\partial E}{\partial T}\right)_V \qquad [7.6]$$

for c_V, and

$$c_p = \frac{(\delta Q)_p}{\mathrm{d}T} = \left(\frac{\partial E}{\partial T}\right)_p + p\left(\frac{\partial V}{\partial T}\right)_p \qquad [7.7]$$

for c_p. If we substitute the expression

$$\left(\frac{\partial E}{\partial T}\right)_p = \left(\frac{\partial E}{\partial T}\right)_V + \left(\frac{\partial E}{\partial V}\right)_T \left(\frac{\partial V}{\partial T}\right)_p$$

into Eq. [7.7], we obtain

$$c_p - c_V = \left[p + \left(\frac{\partial E}{\partial V} \right)_T \right] \left(\frac{\partial V}{\partial T} \right)_p .$$

This formula can be simplified with the aid of the second law. If $E(p, T)$ and $V(p, T)$ are given, then

$$dE = \left(\frac{\partial E}{\partial p} \right)_T dp + \left(\frac{\partial E}{\partial T} \right)_p dT ,$$

$$dV = \left(\frac{\partial V}{\partial p} \right)_T dp + \left(\frac{\partial V}{\partial T} \right)_p dT ,$$

and

$$\delta Q = dE + p\, dV$$
$$= \left[\left(\frac{\partial E}{\partial p} \right)_T + p \left(\frac{\partial V}{\partial p} \right)_T \right] dp + \left[\left(\frac{\partial E}{\partial T} \right)_p + p \left(\frac{\partial V}{\partial T} \right)_p \right] dT ,$$

which gives the formula

$$\delta Q = c_p\, dT + \left[\left(\frac{\partial E}{\partial p} \right)_T + p \left(\frac{\partial V}{\partial p} \right)_T \right] dp . \qquad [7.8]$$

2. *Ideal gases.* Since for ideal gases $(\partial E/\partial V)_T = 0$, it follows that

$$\left(\frac{\partial E}{\partial T} \right)_p = \left(\frac{\partial E}{\partial T} \right)_V = \frac{dE}{dT} = c_V ,$$

and because $pV = RnT$, it follows that

$$c_p - c_V = p \left(\frac{\partial V}{\partial T} \right)_p = p \frac{Rn}{p} = nR .$$

Per mole,

$$c_p - c_V = R .$$

But c_p can be well determined experimentally and, from the speed of sound, so can $\varkappa = c_p/c_V$.[3] From these values we obtain

$$R = 1.985 \ \text{cal/degree} \sim 2 \ \text{cal/degree} ,$$

_____
[3] See p. 15.

which, combined with the previously given value of R, determines the mechanical equivalent of heat as

$$J = 4.19 \times 10^7 \text{ erg/cal.}\,[4]$$

For isothermal changes of state of ideal gases ($dT = 0$), since these changes are also isoenergetic ($dE = 0$), we have

$$\Delta Q = Q_2 - Q_1 = \int_1^2 p\,dV = \int_1^2 RT\frac{dV}{V} = RT\log\frac{V_2}{V_1}.$$

For adiabatic changes of state of ideal gases ($\delta Q = 0$), we have

$$dE + p\,dV = c_V\,dT + p\,dV = 0 \quad \text{or} \quad dT = -\frac{p\,dV}{c_V}.$$

From $pV = RT$ it follows that

$$p\,dV + V\,dp = R\,dT = -\frac{R}{c_V}p\,dV$$

or

$$p\,dV\left(1 + \frac{R}{c_V}\right) + V\,dp = 0,$$

which can be written as

$$\varkappa\frac{dV}{V} = -\frac{dp}{p}. \qquad [7.9]$$

This differential equation has the solution

$$pV^\varkappa = \text{constant}$$

(adiabatic equation of state of an ideal gas).

Since the propagation of sound in a gas consists of adiabatic compressions and rarefactions, $\varkappa$ appears in the speed

---

[4] Translator's Note: In the German edition, this value is incorrectly given as 4.15×10^7 erg/cal.

of sound. The equation for the speed of sound is

$$u^2 = \left(\frac{\mathrm{d}p}{\mathrm{d}\varrho}\right)_{\mathrm{ad}} = -\frac{v^2}{M}\left(\frac{\mathrm{d}p}{\mathrm{d}v}\right)_{\mathrm{ad}}.$$

From this and the adiabatic equation of state, we obtain

$$u^2 = \frac{p}{\varrho}\varkappa = \varkappa\frac{R}{M}T,$$

(M = molar weight, u = speed of sound).

From $pV^\varkappa$ = constant, the following formulas can be derived:

$$p^{\frac{c_V}{R}}V^{\frac{c_p}{R}} = \text{constant}, \qquad \frac{1}{p}T^{\frac{c_p}{R}} = \text{constant}, \qquad T^{\frac{c_V}{R}}V = \text{constant},$$

and

$$\varkappa p\left(\frac{\partial V}{\partial p}\right)_{\mathrm{ad}} + V = 0,$$

which also follows from Eq. [7.9].

c. Equilibrium between two phases

Abbreviations: g = gaseous state, l = liquid state.

Along the equilibrium curve $p = p(T)$, we know from experiment that arbitrary relative amounts of each of two phases can coexist. In equilibrium, $p = p_g = p_l$ and $T = T_g = T_l$.

According to the first law, the value of the heat of transformation (heat of vaporization) is given by

$$\lambda = \Delta Q = (E_g - E_l) + p(V_g - V_l),$$

where the first term is the difference in energy between the gaseous and liquid phases and the second term is the work done during vaporization.

The change $\mathrm{d}\lambda$ in the heat of transformationalong the vaporization curve $p(T)$ amounts to

$$\mathrm{d}\lambda = (\mathrm{d}E + p\,\mathrm{d}V)_g - (\mathrm{d}E + p\,\mathrm{d}V)_l + (V_g - V_l)\,\mathrm{d}p.$$

On the other hand, for the heat added along the vaporization curve, we may write

$$\delta Q = \bar{c}\, \mathrm{d}T\ .$$

If we substitute this in Eq. [7.8] we obtain

$$\bar{c} = c_p + \left[\left(\frac{\partial E}{\partial p}\right)_T + p\left(\frac{\partial V}{\partial p}\right)_T\right]\frac{\mathrm{d}p}{\mathrm{d}T}$$

and

$$\mathrm{d}\lambda = (\bar{c}_g - \bar{c}_l)\,\mathrm{d}T + (V_g - V_l)\frac{\mathrm{d}p}{\mathrm{d}T}\,\mathrm{d}T$$

or

$$\frac{\mathrm{d}\lambda}{\mathrm{d}T} = (\bar{c}_g - \bar{c}_l) + (V_g - V_l)\frac{\mathrm{d}p}{\mathrm{d}T}\ .$$

Formally, we obtain the same relations in the case of melting.

Chapter 2. The Second Law

The second law distinguishes heat from the other forms of energy. It indicates a direction in time and makes apparent that heat is a disordered form of energy.

8. FORMULATIONS OF THE SECOND LAW

a. *Clausius*

There does not exist a device which, working in a cycle, permits heat to be transferred from a reservoir at one temperature to one at a higher temperature without compensating changes (that is, unless at the same time mechanical work is done, or energy is supplied from the surroundings by some other means).

The inverse of the process considered in Clausius's formulation is possible The existence of heat-conducting connections proves this. Clausius says that *heat conduction* is an *irreversible process*.

The term "irreversible" can be defined by means of the above formulation or by the following statement: A process is called *irreversible* if the initial state cannot be reached from the final state without work being done or other changes occurring (that is, without compensation).

A quasi-static process is always reversible.

b. *Thomson*

There does not exist a device which, working in a cycle, permits heat to be removed from a body and transformed

into work unless other changes result (at least in other
bodies).

The inverse of this process is again possible, as internal
friction demonstrates. Thomson says that *internal friction*
is an *irreversible process*. This formulation states that there
exists no device which can transform heat into work without
compensating changes; that is, a perpetual motion machine
of the second kind does not exist.

9. QUANTITATIVE PREDICTIONS OF THE SECOND LAW

a. *The Carnot cycle*

The Carnot cycle is a quasi-static (infinitely slow) process.
It consists of two adiabatic processes ($\delta Q = 0$) and two isothermal processes ($dt = 0$). Since quasi-static processes are
reversible, therefore the Carnot cycle is also reversible.

Isotherms are curves along which the temperature is
constant. They are obtained from isothermal changes of
state. Adiabatic curves are obtained from adiabatic changes
of state. Along adiabatic curves heat is neither added nor
taken away; therefore, $\delta Q = 0$.

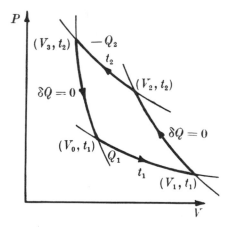

Figure 9.1

The Carnot cycle proceeds as follows: By an isothermal change of state (that is, along the isotherm t_1), we bring the given substance from the state (V_0, t_1) to the state (V_1, t_1), during which the amount of heat Q_1 is added. Then we bring the substance to the state (V_2, t_2) by an adiabatic change of state (along the adiabatic curve). Continuing, we remove the amount of heat Q_2 by proceeding along the isotherm t_2 until we reach the state (V_3, t_2). From there we return to the initial state (V_0, t_1) by an adiabatic change of state. Work is done during all four changes of state. According to the first law,

$$Q_1 - Q_2 = W_1 + W_2 + W_3 + W_4 = W.$$

We distinguish between the following two cases:
1. $Q_1 > Q_2$: More heat is added than taken away, whereby the system does work. During the cycle, heat is thereby converted into work, in the course of which the amount of heat Q_2 drops from t_1 to t_2; that is, heat goes from a warmer reservoir to a cooler one.
2. $Q_1 < Q_2$: More heat is taken awayt han added, whereby work is done on the system from outside. During the cycle, work is thereby converted into heat, in the course of which the amount of heat Q_1 rises from t_1 to t_2, that is, from a cooler reservoir to a warmer one. (See figure.)

We can make the following assertion:

$$Q_1/Q_2 = f(t_1, t_2);$$

that is, Q_1/Q_2 is independent of the nature of the substance. If this assertion were false, then we could construct a perpetual motion machine of the second kind by means of two Carnot cycles. We proceed around a Carnot cycle using substance I and around an identical cycle with substance I' in the opposite direction, so that $Q_1 = Q_1'$; that is, no heat is removed from the reservoir at temperature t_1, because what I takes away I' returns. If it were that

$Q_2 < Q_2'$, then more heat would be taken away from the reservoir at temperature t_2 by I' than would be returned to the reservoir by I.

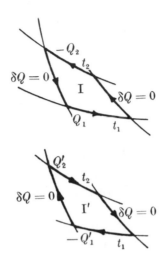

Figure 9.2

According to the first law this difference in heat would be converted into work. Therefore, heat would be taken from the reservoir at temperature t_2 and converted into work without compensating changes, which the second law forbids. If $Q_2 > Q_2'$, then we reverse both processes and obtain the same case as above. Therefore, because of the second law, it must be that

$$Q_2 = Q_2'$$

or

$$\frac{Q_1'}{Q_2'} = \frac{Q_1}{Q_2} = f(t_1, t_2) . \qquad [9.1]$$

If the second law were untrue, that is, if heat conduction were reversible, then work could be obtained from heat without compensating changes. Since it is only possible to convert work into heat without compensating changes, and since it is only possible for heat to go from a higher to a lower temperature without compensating changes, this is the preferred direction in which actual processes take place. We can now also determine which of two temperatures t_1 and t_2 is the higher, namely, that one from which heat can flow to the other without compensating changes [A-3]. Since, according to Section 4, heat which is added is taken as positive, it follows that

$$Q_1 < Q_2 \quad \text{when} \quad t_1 < t_2 .$$

The relation given by Eq. [9.1] can be extended to

$$\frac{Q_1}{Q_2} = \frac{\varphi(t_1)}{\varphi(t_2)}, \qquad [9.2]$$

because

$$\frac{Q_1}{Q_2} = \frac{Q_1}{Q_0}\frac{Q_0}{Q_2} = \frac{Q_1/Q_0}{Q_2/Q_0} = \frac{f(t_1, t_0)}{f(t_2, t_0)} = \frac{\varphi(t_1)}{\varphi(t_2)} .$$

We now define, as the *thermodynamic temperature scale*,

$$\varphi(t) = \text{constant} \times T ,$$

so that we can write Eq. [9.2] as

$$\frac{Q_1}{Q_2} = \frac{T_1}{T_2}$$

or

$$\frac{Q_1}{T_1} = \frac{Q_2}{T_2} . \qquad [9.3]$$

Through this equation, the temperature is defined to within a constant factor; the zero point is now fixed. That the temperature T is positive is contained in the second law [A-6]. We have also defined what it means to say that one temperature is twice as large as another.

b. Arbitrary quasi-static processes

An arbitrary quasi-static cyclic process can be reduced to changes in a single heat reservoir with the help of auxiliary—in the limiting case, infinitely many—Carnot cycles. First, we make a decomposition into only a finite number of individual processes with temperatures T_k, at which the quantities of heat δQ_k are added $(k = 1, 2, ..., Z)$. The quantities of heat δQ_{0k} are removed from a single reservoir at temperature T_0 with the aid of Carnot cycles. We have

$$\frac{\delta Q_k}{T_k} = \frac{\delta Q_{0k}}{T_0}$$

for all k. For the cycle we then obtain

$$\sum_k \frac{\delta Q_k}{T_k} = \sum_k \frac{\delta Q_{0k}}{T_0} = \frac{1}{T_0} \sum_k \delta Q_{0k} = \frac{Q_0}{T_0}.$$

Since we have completed a cycle, and since no changes result, heat could neither have been added to nor taken away from the heat reservoir at T_0. Thus, it must be that $Q_0 = 0$ [A-7], or

$$\sum_k \frac{\delta Q_k}{T_k} = 0$$

and, in the limiting case,

$$\lim_{z \to \infty} \sum_{k=1}^{z} \frac{\delta Q_k}{T_k} = \oint \frac{\delta Q}{T} = 0.$$

Accordingly, $\int_1^2 \delta Q/T$ is a function of state (independent of path). If we introduce a reference state o, we can define *entropy* as

$$S_n = \int_o^n \frac{\delta Q}{T}. \qquad [9.4]$$

For quasi-static changes of state we have

$$\int_1^2 \frac{\delta Q}{T} = S_2 - S_1 ;$$

and for cyclic processes we have

$$\Delta S = \oint \frac{\delta Q}{T} = 0 \ .$$

c. Rapidly occurring processes

It still remains for us to investigate how the entropy changes during rapidly occurring (irreversible) processes. To that end we consider a process in a *closed box*, in which the *internal energy remains constant*. Thermodynamics re-

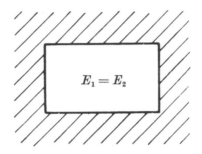

Figure 9.3

quires that the initial state be attainable from the final state along a quasi-static path. The change in entropy along this quasi-static path is

$$S_1 - S_2 = \int_2^1 \frac{\delta Q}{T} = \frac{Q_0}{T_0} \ .$$

Here Q_0 is the heat removed from a reservoir at temperature T_0, and $Q_0 < 0$. Were $Q_0 > 0$, then heat would have been converted into work without compensating changes, which contradicts the second law of thermodynamics. Thus, $S_1 < S_2$; that is, the *entropy has increased*.

Concerning the second law, the following is to be noted: The thermodynamic function of state can also be determined without using irreversible processes, on which the statements of Clausius and Thomson are based. This is done by weakening the principles of impossibility somewhat by only stating something about quasi-static processes. In a quasi-static cyclic process:

1. Heat cannot be converted into work unless, at the same time, a corresponding quantity of heat is taken from a warmer reservoir to a colder one;
2. Heat cannot go from a colder reservoir to a warmer one unless, at the same time, a corresponding amount of work goes into heat;
3. Work cannot be converted into heat unless, at the same time, a corresponding quantity of heat is taken from a colder reservoir to a warmer one;
4. Heat cannot go from a warmer reservoir to a colder one unless, at the same time, a corresponding quantity of heat is converted into work.

If we postulate the validity of the above four axioms for quasi-static changes of state, then an entropy function exists. If we require that only axioms 1 and 2 are true for rapidly occurring processes, then an entropy function exists nevertheless. In a closed system, the entropy can change only in one direction; it can either increase or it can decrease. However, both possibilities are never simultaneously available [A-8].

A further remark concerns the question of whether the temperature defined by the Carnot processes must necessarily be positive [A-6]. [The same question occurs for the sign of the integrating factor in the formulation of thermodynamics by Carathéodory (Section 11).] From the axiomatic viewpoint, Mrs. T. Ehrenfest-Afanasjewa showed that this does not follow solely from the remaining assumptions [*Z. Physik* **33**, 933 (1925) and a correction in *Z. Physik* **34**, 638 (1925)]. However, in order that this question not be a purely formal one, the statistical methods (canonical ensemble) which form the subject matter of the following set of lectures must be applied here. From this broadened viewpoint N. F. Ramsey showed that special systems can be characterized by a negative temperature under certain circumstances [*Phys. Rev.* **103**, 20 (1956)].

10. MATHEMATICAL FORMULATION OF THE SECOND LAW

The mathematical statement of the second law is that $\delta Q/T = \mathrm{d}S$ is an *exact differential*.

This statement is no longer tied to quasi-static processes. It defines entropy in complete generality. We have

$$\mathrm{d}S = \frac{\mathrm{d}E + \delta W}{T} = \frac{\mathrm{d}E + \sum_{k=1}^{n+1} y_k\,\mathrm{d}x_k}{T},$$

where [A-2] $E = E(x_1, x_2, \ldots, x_{n+1})$, and $y_{n+1} = 0$. If we introduce the functions

$$w_k = \frac{1}{T}\left(\frac{\partial E}{\partial x_k} + y_k\right) \qquad (k = 1, 2, \ldots, n)$$

and

$$w_{n+1} = \frac{1}{T}\left(\frac{\partial E}{\partial T}\right),$$

then we can write

$$\mathrm{d}S = \sum_{k=1}^{n+1} w_k\,\mathrm{d}x_k \qquad \text{(differential form of Pfaff)}.$$

Since $\mathrm{d}S$ is an exact differential, it is necessary that [A-2]

$$\frac{\partial w_i}{\partial x_k} = \frac{\partial w_k}{\partial x_i} \qquad (i, k = 1, 2, 3, \ldots, n+1).$$

Example: For homogeneous substances, we have $\mathrm{d}S = (\mathrm{d}E + p\,\mathrm{d}V)/T$. If we choose E, V as independent variables, that is, $T = T(E, V)$ and $p = p(E, V)$, then

$$\left[\frac{\partial(1/T)}{\partial V}\right]_E = \left[\frac{\partial(p/T)}{\partial E}\right]_V.$$

However, this notation is very inexpedient. Therefore, we shall introduce other functions with more convenient independent variables later.

11. AXIOMATIC FOUNDATION OF THERMODYNAMICS FOLLOWING CARATHÉODORY [1]

a. Pfaff's linear differential forms

Let the functions $X_i(x_1, x_2, ..., x_n)$ be given. In order that the expression (Pfaff's form)

$$\delta Q = \sum_{k=1}^{n} X_k(x_1, x_2, ..., x_n)\, \mathrm{d}x_k$$

be the exact differential $\mathrm{d}f$ of a function $f(x_1, x_2, ..., x_n)$, the conditions

$$\frac{\partial X_i}{\partial x_k} - \frac{\partial X_k}{\partial x_i} = 0$$

must be fulfilled for all i and k, as a consequence of

$$\frac{\partial f}{\partial x_k} = X_k \qquad \text{and} \qquad \frac{\partial X_k}{\partial x_i} = \frac{\partial^2 f}{\partial x_i \partial x_k} = \frac{\partial^2 f}{\partial x_k \partial x_i} = \frac{\partial X_i}{\partial x_k}.$$

If δQ is not an exact differential, then the case when a function $\tau(x_1, x_2, ..., x_n)$ (the reciprocal of which is called an *integrating factor*) can be introduced to make

$$\frac{\delta Q}{\tau} = \sum_{k} \frac{X_k}{\tau}\, \mathrm{d}x_k = \mathrm{d}f$$

an exact differential is of especial interest in thermodynamics. The conditions for the existence of such an integrating factor $1/\tau$ are given by

$$\frac{\partial(X_i/\tau)}{\partial x_k} - \frac{\partial(X_k/\tau)}{\partial x_i} = 0 \qquad i, k = 1, 2, ..., n.$$

[1] References for Section 11: C. CARATHÉODORY, *Math. Ann.* **67**, 355 (1909), and *Sitzber. preuss. Akad. Wiss., Physik.-math. Kl.*, Jahrband 1925, p. 39.

The function τ can be eliminated by forming the sum

$$(ikl) + (kli) + (lik) = 0 ,$$

where, for example,

$$(ikl) = \tau X_l \left[\frac{\partial (X_i/\tau)}{\partial x_k} - \frac{\partial (X_k/\tau)}{\partial x_i} \right] = 0 .$$

We obtain

$$X_l \left(\frac{\partial X_i}{\partial x_k} - \frac{\partial X_k}{\partial x_i} \right) + X_i \left(\frac{\partial X_k}{\partial x_l} - \frac{\partial X_l}{\partial x_k} \right) + X_k \left(\frac{\partial X_l}{\partial x_i} - \frac{\partial X_i}{\partial x_l} \right) = 0 .$$

This condition must be fulfilled for all possible triples (i, k, l). It can be shown that these conditions for the existence of an integrating factor $1/\tau$ are also sufficient. For two variables this condition is not applicable. Now, we want to consider those curves $\delta Q = \sum_k X_k \, dx_k$, which pass through a fixed point. Two cases must be distinguished:

1. If there exists an integrating factor $(\tau \neq 0)$, then it follows from $\delta Q = \tau \, df$ that, when $\delta Q = 0$, then $df = 0$ also; this implies that $f(x_1, x_2, ..., x_n) = $ constant. Accordingly, all curves $\delta Q = 0$ lie in an $(n-1)$-dimensional hyperplane through the fixed point. Thus, in the neighborhood of the fixed point there exist points arbitrarily close which do not lie in the hyperplane and, therefore, cannot be reached from the fixed point along curves $\delta Q = 0$. Since the fixed point may be chosen arbitrarily, the following statement is true: If an integrating factor exists $(\tau \neq 0)$, then, for every point x in the n-dimensional space, there are points arbitrarily close to x which cannot be reached along curves $\delta Q = 0$.

2. If there does not exist an integrating factor, then all neighboring points can be reached along curves $\delta Q = 0$.

In order to prove this statement, we use the fact that the Pfaff linear differential forms can be brought into the normal form

$$\delta Q = x_2 \, dx_1 + x_4 \, dx_3 + x_6 \, dx_5 + ... + x_{2n} \, dx_{2n-1} + k \, dx_{2n+1}$$

if the number of variables is odd $(2n+1)$, and

$$\delta Q = x_2\,dx_1 + x_4\,dx_3 + x_6\,dx_5 + \ldots + x_{2n}\,dx_{2n-1}$$

if the number of variables is even $(2n)$ (that is, here k is zero). We will carry out the proof for $(2n+1)$ variables.

Let $P^0(x_1^0, x_2^0, \ldots, x_{2n+1}^0)$ be the fixed point and $P'(x_1', x_2', \ldots, x_{2n+1}')$ the point to be reached along curves $\delta Q = 0$. Let the running coordinates be $x_1, x_2, x_3, \ldots, x_{2n+1}$. First, we let $x_1, x_3, x_5, \ldots, x_{2n+1}$ be constant, that is, $x_1 = x_1^0$, $x_3 = x_3^0$, $x_5 = x_5^0, \ldots, x_{2n+1} = x_{2n+1}^0$ (therefore $\delta Q = 0$), and go from P^0 to the point $\overline{P}(x_1^0, \xi_2, x_3^0, \xi_4, \ldots, \xi_{2n}, x_{2n+1}^0)$, whose coordinates $\xi_2, \xi_4, \xi_6, \ldots, \xi_{2n}$ satisfy the condition

$$\xi_2(x_1' - x_1^0) + \xi_4(x_3' - x_3^0) + \ldots$$
$$+ \xi_{2n}(x_{2n-1}' - x_{2n-1}^0) + k(x_{2n+1}' - x_{2n+1}^0) = 0\,.$$

Now, we leave $x_2, x_4, \ldots, x_{2n}$ constant, that is, $x_2 = \xi_2$, $x_4 = \xi_4$, $x_6 = \xi_6, \ldots, x_{2n} = \xi_{2n}$, and go along the surface defined by

$$\xi_2(x_1 - x_1^0) + \xi_4(x_3 - x_3^0) + \ldots$$
$$+ \xi_{2n}(x_{2n-1} - x_{2n-1}^0) + k(x_{2n+1} - x_{2n+1}^0) = 0$$

from $\overline{P}$ to $P^*(x_1', \xi_2, x_3', \xi_4, \ldots, \xi_{2n}, x_{2n+1}')$, in which case again $\delta Q = 0$. Lastly, we again leave $x_1, x_3, x_5, \ldots, x_{2n+1}$ constant, that is, $x_1 = x_1'$, $x_3 = x_3'$, $x_5 = x_5', \ldots, x_{2n+1} = x_{2n+1}'$, and go from P^* to $P'(x_1', x_2', x_3', \ldots, x_{2n+1}')$, in which case $\delta Q = 0$ also. Consequently, P' can be reached from P^0 along curves $\delta Q = 0$.

If everywhere in the above proof x_{2n+1} is left out and k is set equal to zero, we obtain the analogous proof for $2n$ variables.

b. Application to thermodynamics

We need the following concepts:

1. *Adiabatic walls*: Changes of state are possible only through mechanical means. *Adiabatic walls do not conduct heat.*

2. *Heat conducting walls*: All nonadiabatic walls are heat conducting.

3. *Quasi-static changes of state*: These changes of state are very slow, infinitely slow in the limiting case, so that the intermediate states form a continuous sequence of equilibrium states.

4. *Nonequilibrium changes of state*: All non-quasi-static changes of state are nonequilibrium (rapidly occurring).

The quasi-static changes of state of a closed system within adiabatic walls, that is, the adiabatic quasi-static processes, lead to the existence of adiabatic curves, along which $\delta Q = 0$. For quasi-static changes of state within heat-conducting walls, that is, for processes involving heat conduction, there is thermal equilibrium between the inside and outside. If two substances are separated by a heat-conducting wall, then there exists a relation [A-2]

$$F(p, V, \overline{p}, \overline{V}) = 0 \,.$$

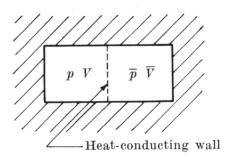

Heat-conducting wall

Figure 11.1

This relation must have the form $t(p, V) = \overline{t}(\overline{p}, \overline{V})$ for the following reason: If (p_1, V_1) and (p_2, V_2) are different states of one substance, and if $(\overline{p}_1, \overline{V}_1)$ and $(\overline{p}_2, \overline{V}_2)$ are different states of the other substance, then experiment shows that the three conditions [A-2]

$$\left.\begin{array}{l} F(p_1, V_1, \overline{p}_1, \overline{V}_1) = 0 \\[4pt] F(p_2, V_2, \overline{p}_1, \overline{V}_1) = 0 \\[4pt] F(p_1, V_1, \overline{p}_2, \overline{V}_2) = 0 \end{array}\right\} \quad \text{imply} \quad F(p_2, V_2, \overline{p}_2, \overline{V}_2) = 0 \,.$$

However, this is possible only if the relation $F = 0$ has the form $t(p, V) - \bar{t}(\bar{p}, \bar{V}) = 0$.

In order to formulate the first law, we can now restrict ourselves to adiabatic but not necessarily quasi-static changes of state. For such a process we define

$$E_2 - E_1 = -\int_1^2 \delta W \,.$$

Here, it is to be noted that there always exists an adiabatic [A-9] path which either leads from 1 to 2 or from 2 to 1. For adiabatic [A-9] changes of state, we have

$$\delta W = \sum_k y_k \, dx_k \qquad \text{and} \qquad dE + \delta W = 0 \,,$$

for homogeneous substances $dE + p \, dV = 0$. Therefore, the equation for adiabatic changes of a homogeneous substance is

$$\left(\frac{\partial E}{\partial t}\right) dt + \left[\left(\frac{\partial E}{\partial V}\right) + p\right] dV = 0 \,,$$

where the temperature function $t(p, V)$ is still completely arbitrary and need not be monotonic. For two homogeneous substances which are separated by a heat conducting wall, that is, which are in thermal equilibrium, the equation for adiabatic changes is

$$\left[\left(\frac{\partial \bar{E}}{\partial t}\right) + \left(\frac{\partial E}{\partial t}\right)\right] dt + \left[\left(\frac{\partial \bar{E}}{\partial \bar{V}}\right) + \bar{p}\right] d\bar{V} + \left[\left(\frac{\partial E}{\partial V}\right) + p\right] dV = 0 \,.$$

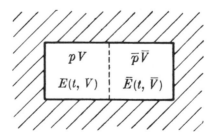

Figure 11.2

For nonadiabatic changes of state, that is, changes involving heat conduction, we define

$$\int_1^2 \delta Q = E_2 - E_1 + \int_1^2 \delta W \,.$$

This definition determines the unit of heat. If the temperature is held fixed, $t(p, V) = $ constant, then we obtain the equation of the isotherms. Nonadiabatic changes of state can always be traced back to heat conduction.

For the second law, Carathéodory replaces the formulations of Clausius and Thomson with the following requirement, which at first shall be valid only for quasi-static changes of state: For arbitrary initial states, there exist neighboring states which cannot be reached from the initial state by means of quasi-static adiabatic changes ($\delta Q = 0$). We can also say that there exist neighboring states which cannot be reached from the initial state along adiabatic curves. Thus, according to Carathéodory, the existence of states which cannot be reached adiabatically implies the existence of an integrating factor. For a single homogeneous substance, the equation for adiabatic changes is

$$\left(\frac{\partial E}{\partial t}\right) dt + \left[\left(\frac{\partial E}{\partial V}\right) + p\right] dV = 0 \,.$$

Since a linear differential form with two variables can always be written with an integrating factor, it follows that

$$\tau \, d\sigma = \left(\frac{\partial E}{\partial t}\right) dt + \left[\left(\frac{\partial E}{\partial V}\right) + p\right] dV = 0 \,,$$

where $\tau = \tau(t, V)$ and $\sigma = \sigma(t, V)$.

For two homogeneous substances which are in thermal equilibrium, the equation for adiabatic changes of the combined system is, as given above,

$$\left[\left(\frac{\partial E_1}{\partial t}\right) + \left(\frac{\partial E_2}{\partial t}\right)\right] dt + \left[\left(\frac{\partial E_1}{\partial V_1}\right) + p_1\right] dV_1 + \left[\left(\frac{\partial E_2}{\partial V_2}\right) + p_2\right] dV_2 = 0 \,.$$

Since the integrating factors for the individual substances exist, we have

$$\tau_1(t, V_1)\, \mathrm{d}\sigma_1(t, V_1) = \left(\frac{\partial E_1}{\partial t}\right) \mathrm{d}t + \left[\left(\frac{\partial E_1}{\partial V_1}\right) + p_1\right] \mathrm{d}V_1,$$

$$\tau_2(t, V_2)\, \mathrm{d}\sigma_2(t, V_2) = \left(\frac{\partial E_2}{\partial t}\right) \mathrm{d}t + \left[\left(\frac{\partial E_2}{\partial V_2}\right) + p_2\right] \mathrm{d}V_2\ .$$

However, according to Carathéodory, an integrating factor for the composite system also must exist; that is,

$$\tau(t, V_1, V_2)\, \mathrm{d}\sigma(t, V_1, V_2)$$

$$= \left[\left(\frac{\partial E_1}{\partial t}\right) + \left(\frac{\partial E_2}{\partial t}\right)\right] \mathrm{d}t + \left[\left(\frac{\partial E_1}{\partial V_1}\right) + p_1\right] \mathrm{d}V_1 + \left[\left(\frac{\partial E_2}{\partial V_2}\right) + p_2\right] \mathrm{d}V_2\ ,$$

from which it follows that

$$\tau\, \mathrm{d}\sigma = \tau_1\, \mathrm{d}\sigma_1 + \tau_2\, \mathrm{d}\sigma_2\ . \qquad [11.1]$$

By introducing new variables t, σ_1, and σ_2 instead of the variables t, V_1, and V_2, we find

$$\frac{\partial \sigma}{\partial \sigma_1} = \frac{\tau_1}{\tau}, \qquad \frac{\partial \sigma}{\partial \sigma_2} = \frac{\tau_2}{\tau}, \qquad \frac{\partial \sigma}{\partial t} = 0\ .$$

From this we see that σ is independent of t and, further, that τ_1/τ and τ_2/τ are likewise independent of t. Therefore, we can write

$$\tau_1 = f(t)\Sigma_1(\sigma_1)\ ,$$

$$\tau_2 = f(t)\Sigma_2(\sigma_2)\ ,$$

$$\tau = f(t)\Sigma(\sigma_1, \sigma_2)\ . \qquad [11.2]$$

It remains only to show that $\Sigma(\sigma_1, \sigma_2) = \Sigma(\sigma)$. If we substitute the expressions given by Eq. [11.2] into Eq. [11.1], we obtain $\Sigma \mathrm{d}\sigma = \Sigma_1 \mathrm{d}\sigma_1 + \Sigma_2 \mathrm{d}\sigma_2$, from which it follows that

$$(\partial\sigma/\partial\sigma_1)\Sigma = \Sigma_1 \quad \text{and} \quad (\partial\sigma/\partial\sigma_2)\Sigma = \Sigma_2\ .$$

If we now differentiate with respect to σ_2 and σ_1, respectively, we obtain

$$\frac{\partial \sigma}{\partial \sigma_1} \frac{\partial \Sigma}{\partial \sigma_2} + \frac{\partial^2 \sigma}{\partial \sigma_1 \partial \sigma_2} \Sigma = 0 = \frac{\partial \sigma}{\partial \sigma_2} \frac{\partial \Sigma}{\partial \sigma_1} + \frac{\partial^2 \sigma}{\partial \sigma_2 \partial \sigma_1} \Sigma$$

or

$$\frac{\partial \sigma}{\partial \sigma_1} \frac{\partial \Sigma}{\partial \sigma_2} - \frac{\partial \sigma}{\partial \sigma_2} \frac{\partial \Sigma}{\partial \sigma_1} = 0 .$$

This quantity is exactly the Jacobian, and the fact that it equals zero implies

$$\Sigma(\sigma_1, \sigma_2) = \Sigma(\sigma) .$$

We now define the thermodynamic temperature, to within a constant multiplicative factor [A-6], as the reciprocal of the integrating factor:

$$f(t) = T .$$

In addition, we make the following further definitions:

$$dS = \Sigma(\sigma) \, d\sigma , \quad dS_1 = \Sigma_1(\sigma_1) \, d\sigma_1 , \quad dS_2 = \Sigma_2(\sigma_2) \, d\sigma_2 ,$$

or

$$S = \int \Sigma(\sigma) \, d\sigma + \text{constant},$$

$$S_1 = \int \Sigma_1(\sigma_1) \, d\sigma_1 + \text{constant},$$

$$S_2 = \int \Sigma_2(\sigma_2) \, d\sigma_2 + \text{constant} .$$

We then have

$$S = S_1 + S_2 \quad \text{and} \quad \delta Q = T \, dS = T \, dS_1 + T \, dS_2 .$$

These two expressions are exactly equivalent to the second law. We can thus say that if two systems are in thermodynamic equilibrium, then the second law requires the existence of an integrating factor.

If the existence of adiabatically unattainable states is required also for rapidly occurring adiabatic changes of

state, then the following statement is valid: For all possible changes of state (rapid and quasi-static) in a closed
system the entropy can only increase or decrease; both possibilities are never simultaneously available [A-8]. Let the
initial state be specified by (V_1^0, V_2^0, S^0) and the final state
by (V_1, V_2, S). We introduce an intermediate state, (V_1, V_2, S^0),
which is quasi-statically attainable (since $\Delta S = 0$). Without
changing the volumes V_1 and V_2 we carry out a rapidly
occurring process (stirring or rubbing); that is, $(V_1, V_2,$
$S^0) \to (V_1, V_2, S)$. If the entropy could both increase or decrease as a result of a rapid process, then all states would
be attainable through rapid processes. However, there must
be a restriction. The only possible one consists in allowing
only one algebraic sign for entropy changes; this means
that the entropy can either increase or decrease, but not
both. The conventional normalization is that the entropy
can only increase in a closed system [A-8].

12. FREE ENERGY

a. Definition

We define the free energy F as

$$F = E - TS .\qquad [12.1]$$

It follows from this definition that

$$dF = dE - T\,dS - S\,dT$$

and, if the second law is introduced, that [A-2]

$$dF = - S\,dT - \delta W = - S\,dT - \sum_{k=1}^{n+1} y_k\,dx_k , \qquad y_{n+1} = 0 . \quad [12.2]$$

Therefore, $F = F(x_1, x_2, \dots, x_n, T)$. Further, we have the
relations

$$\left(\frac{\partial F}{\partial x_k}\right)_T = - y_k \quad \text{and} \quad \left(\frac{\partial F}{\partial T}\right)_x = - S .$$

If we substitute these relations into Eq. [12.1], then we

obtain

$$F = E + T \left(\frac{\partial F}{\partial T} \right)_x$$

or

$$E = F - T \left(\frac{\partial F}{\partial T} \right)_x = - T^2 \frac{\partial}{\partial T} \left(\frac{F}{T} \right)_x. \qquad [12.3]$$

The function F is used when the temperature T appears as an independent variable, and the entropy S is used when the internal energy E appears as an independent variable.

b. *Application to homogeneous substances*

Assume that $n = 1$, $x = V$, $y = p$, $F = F(V, T)$. Then,

$$dF = - p \, dV - S \, dT, \qquad [12.4]$$

$$\left(\frac{\partial F}{\partial V} \right)_T = - p, \quad \text{and} \quad \left(\frac{\partial F}{\partial T} \right)_V = - S.$$

Because F is a function of state, we have

$$\left(\frac{\partial^2 F}{\partial T \partial V} \right) = - \left(\frac{\partial p}{\partial T} \right)_V = - \left(\frac{\partial S}{\partial V} \right)_T. \qquad [12.5]$$

This identity is very important for homogeneous substances. From Eqs. [12.3] and [12.5], we have [A-5]

$$\left(\frac{\partial E}{\partial V} \right)_T = \left(\frac{\partial F}{\partial V} \right)_T - T \left(\frac{\partial^2 F}{\partial T \partial V} \right) = - p + T \left(\frac{\partial S}{\partial V} \right)_T. \qquad [12.6]$$

For the specific heat at constant volume, c_V, we have [A-5]

$$c_V = \frac{(\delta Q)_V}{dT} = T \left(\frac{\partial S}{\partial T} \right)_V = - T \left(\frac{\partial^2 F}{\partial T^2} \right)_V,$$

and for the specific heat at constant pressure, c_p, we have

$$c_p = \frac{(\delta Q)_p}{dT} = T \left(\frac{\partial S}{\partial T} \right)_p = T \left\{ \left(\frac{\partial S}{\partial T} \right)_V + \left(\frac{\partial S}{\partial V} \right)_T \left(\frac{\partial V}{\partial T} \right)_p \right\};$$

from these we obtain

$$c_p - c_V = T \left(\frac{\partial S}{\partial V} \right)_T \left(\frac{\partial V}{\partial T} \right)_p = T \left(\frac{\partial p}{\partial T} \right)_V \left(\frac{\partial V}{\partial T} \right)_p. \qquad [12.7]$$

It must be noted that all of these relations are valid only for homogeneous substances.

From the first law we obtained

$$c_p - c_V = \left[\left(\frac{\partial E}{\partial V}\right)_T + p\right]\left(\frac{\partial V}{\partial T}\right)_p .$$

If this formula is combined with Eq. [12.6], then we obtain exactly Eq. [12.7]. Further, we deduce from the second law that

$$\left(\frac{\partial c_V}{\partial V}\right)_T = T\left(\frac{\partial^2 p}{\partial T^2}\right)_V .$$

Equation [12.7] can be further transformed mathematically. Because of the existence of the equation of state, we can assume that $V(p, T)$ is given. Then,

$$dV = \left(\frac{\partial V}{\partial p}\right)_T dp + \left(\frac{\partial V}{\partial T}\right)_p dT .$$

At constant volume we have $(\partial V/\partial T)_V = 0$. Therefore,

$$\left(\frac{\partial V}{\partial T}\right)_V = \left(\frac{\partial V}{\partial p}\right)_T \left(\frac{\partial p}{\partial T}\right)_V + \left(\frac{\partial V}{\partial T}\right)_p = 0$$

or

$$\left(\frac{\partial p}{\partial T}\right)_V = -\frac{(\partial V/\partial T)_p}{(\partial V/\partial p)_T} .$$

Therefore, Eq. [12.7] can be written

$$c_p - c_V = -\frac{T[(\partial V/\partial T)_p]^2}{(\partial V/\partial p)_T} .$$

13. GIBBS'S FUNCTION

a. Definition

For the free energy F the independent variables are $x_1, x_2, x_3, \ldots, x_n, T$. On the other hand, the Gibbs function Φ is to be a *thermodynamic function* whose independent variables are $y_1, y_2, y_3, \ldots, y_n, T$. Therefore, we apply the following Legendre transformation to F:

$$\Phi = F - \sum_{k=1}^{n} x_k \left(\frac{\partial F}{\partial x_k}\right)_T$$

$$= F + \sum_{k=1}^{n} x_k y_k = E - TS + \sum_{k=1}^{n} x_k y_k \,, \qquad [13.1]$$

$$d\Phi = dF + \delta W + \sum_{k=1}^{n} x_k \, dy_k$$

$$= - S \, dT + \sum_{k=1}^{n} x_k \, dy_k = \left(\frac{\partial F}{\partial T}\right)_{x_k} dT + \sum_{k=1}^{n} x_k \, dy_k \,. \qquad [13.2]$$

The terms containing dx_k cancel and, therefore, we have $\Phi = \Phi(y_1, y_2, y_3, ..., y_n, T)$. Furthermore,

$$\left(\frac{\partial \Phi}{\partial T}\right)_{y_k} = \left(\frac{\partial F}{\partial T}\right)_{x_k} = - S \qquad \text{and} \qquad \left(\frac{\partial \Phi}{\partial y_k}\right)_T = x_k \,. \qquad [13.3]$$

b. Application to homogeneous substances

We have

$$\Phi = F + pV \qquad \text{and} \qquad d\Phi = - S \, dT + V \, dp \,.$$

Therefore, $\Phi = \Phi(p, T)$. For quasi-static changes of state at constant pressure p and constant temperature T (isothermal isobaric processes), the Gibbs function Φ is constant. Furthermore,

$$\left(\frac{\partial \Phi}{\partial T}\right)_p = - S \qquad \text{and} \qquad \left(\frac{\partial \Phi}{\partial p}\right)_T = V \,.$$

In the previous section we obtained for the specific heat at constant volume [A-5]

$$c_V = - T \left(\frac{\partial^2 F}{\partial T^2}\right)_V \qquad \text{and} \qquad \left(\frac{\partial c_V}{\partial V}\right)_T = T \left(\frac{\partial^2 p}{\partial T^2}\right)_V \,.$$

Analogously, we obtain for the specific heat at constant

pressure

$$c_p = T\left(\frac{\partial S}{\partial T}\right)_p = -T\left(\frac{\partial^2 \Phi}{\partial T^2}\right)_p \quad \text{and} \quad \left(\frac{\partial c_p}{\partial p}\right)_T = -T\left(\frac{\partial^2 V}{\partial T^2}\right)_p.$$

c. Application to ideal gases

With the help of the second law, the three axiomatic properties of ideal gases can be reduced to two. At the same time the agreement of the thermodynamic temperature scale with the gas scale can be demonstrated.

1. Since $(\partial E/\partial V)_T = 0$, it follows that $T(\partial p/\partial T)_V = p$ or $p = f(V)T$. Therefore, p is a linear function of T.
2. The isotherms are $pV =$ constant.

From 1 and 2 follows $pV = (\text{constant}) \times T$ or $pv = RT$ per mole ($v =$ molar volume). That the value of the constant which appears in this equation is R does not follow from thermodynamics. The same gas equation is obtained when the thermodynamic temperature scale is used as when the gas scale is used. Therefore, *the thermodynamic temperature scale agrees with the absolute temperature scale.*

The entropy of an ideal gas is

$$dS = \frac{1}{T}(dE + p\,dV) = c_v \frac{dT}{T} + R\frac{dV}{V},$$

$$S = \int c_v \frac{dT}{T} + R\log V + \text{constant}.$$

For ideal gases $c_v(T) = $ constant, which is not true for arbitrary substances. Therefore, per mole,

$$S = c_v \log T + R\log v + \text{constant}.$$

As long as the quantity of material in the system remains constant (that is, as long as no chemical changes take place), we can accept as part of the definition of the entropy S of arbitrary substances that the entropy of a system increases in proportion to the size of the system if the density is constant. Therefore, we have the following homogeneity property:

$$S(n, V, T) = nS(1, v, T) = nS\left(\frac{V}{n}, T\right)$$

($v=$ molar volume, $n = V/v =$ number of moles). For ideal gases we then have

$$S = n\left(c_v \log T + R \log \frac{V}{n} + a\right).$$

We shall normalize the entropy constant a later.

Because $S = -(\partial F/\partial T)_v$, we have for ideal gases

$$F = n\left(c_v T - c_v T \log T - RT \log \frac{V}{n} - aT + E_0\right),$$

where E_0 is the internal energy per mole extrapolated to $T = 0$, since the energy is independent of the volume. Therefore, because $F = E - TS$, we have

$$E = n(c_v T + E_0),$$

for which result we have assumed that $\partial c_v/\partial T = 0$. The Gibbs function, written in terms of inexpedient variables, is

$$\Phi = F + pV = n\left(c_v T - c_v T \log T - RT \log \frac{V}{n} - aT + E_0 + RT\right).$$

Since $c_v + R = c_p$ and $V/n = RT/p$, we have

$$-RT \log \frac{V}{n} = -RT (\log T + \log R - \log p)$$

and

$$\begin{aligned}\Phi = n[&RT \log p + (c_v + R)T \\ &- (c_v + R)T \log T - aT - RT \log R + E_0] \\ = n(&RT \log p - c_p T \log T - iRT + E_0),\end{aligned}$$

where

$$i = \frac{a - c_p}{R} + \log R = \gamma + \log R ;$$

γ is called the chemical constant.

For ideal gases we therefore have the following relations:

1. *Internal energy:*

$$E = n(c_v T + E_0) ;$$

2. *Entropy:*

$$S = n\left(c_v \log T + R \log \frac{V}{n} + a\right) ;$$

3. *Free energy:*

$$F = n\left(c_v T - c_v T \log T - RT \log \frac{V}{n} - aT + E_0\right) ;$$

4. *Gibbs's function:*

$$\Phi = n(RT \log p - c_v T \log T - iRT + E_0) .$$

The constant a cannot be determined and will be normalized later. (See the section on Nernst's theorem.)

d. Joule–Thomson experiment

A gas flows from a container (1) into another container (2). The entire system is thermally insulated from the outside.

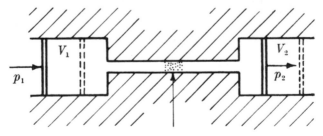

Throttle (e.g., cotton plug)

Figure 13.1

However, an exchange of heat between the throttle and the flowing gas is possible. The work done is

$$W = p_2 V_2 - p_1 V_1 .$$

According to the first law we have

$$p_2 V_2 - p_1 V_1 = E_1 - E_2 ,$$

or

$$E_1 + p_1 V_1 = E_2 + p_2 V_2 = \text{constant} .$$

(Of course, for ideal gases we have $E_1 = E_2$ and $T_1 = T_2$.) According to Eqs. [13.1] and [13.3],

$$E + pV = \Phi - T \left(\frac{\partial \Phi}{\partial T} \right)_p = \text{constant} ,$$

from which we obtain by differentiation

$$- T \left(\frac{\partial^2 \Phi}{\partial T^2} \right)_p \Delta T + \left[\left(\frac{\partial \Phi}{\partial p} \right)_T - T \left(\frac{\partial^2 \Phi}{\partial p \partial T} \right) \right] \Delta p = 0 .$$

Therefore,

$$n c_p \Delta T = \left[T \left(\frac{\partial V}{\partial T} \right)_p - V \right] \Delta p = T^2 \frac{\partial}{\partial T} \left(\frac{V}{T} \right)_p \Delta p .$$

Usually Δp is given empirically. For ideal gases, because $V/T = nR/p$, we have $(\partial/\partial T)(V/T)_p = 0$. In general, $(\partial/\partial T)(V/T)_p$ can have either sign. The temperature at which this expression is zero is called the *inversion point*. The above formula finds its practical application in the Linde air liquefaction process.

e. Equilibrium between two phases A and B

Since pressure and temperature are constant during melting and evaporation processes, the Gibbs function Φ will be used. We have $p_A = p_B$ and $T_A = T_B$, from which it follows that $\Phi_A = \Phi_B$. As has already been shown,[2]

$$\lambda = (E_B - E_A) + p(V_B - V_A) .$$

[2] See p. 15.

Since the process is isothermal, we have from the second law that

$$T(S_B - S_A) = (E_B - E_A) + p(V_B - V_A) = \lambda. \quad [13.4]$$

The derivative along the vapor pressure curve $p(T)$ is given by

$$\left(\frac{\mathrm{d}}{\mathrm{d}T}\right) = \left(\frac{\partial}{\partial T}\right)_p + \left(\frac{\partial}{\partial p}\right)_T \frac{\mathrm{d}p}{\mathrm{d}T}.$$

Since $\Phi_A = \Phi_B$, we have

$$\left(\frac{\partial \Phi_A}{\partial T}\right)_p + \left(\frac{\partial \Phi_A}{\partial p}\right)_T \frac{\mathrm{d}p}{\mathrm{d}T} = \left(\frac{\partial \Phi_B}{\partial T}\right)_p + \left(\frac{\partial \Phi_B}{\partial p}\right)_T \frac{\mathrm{d}p}{\mathrm{d}T}$$

and hence

$$(S_B - S_A) = (V_B - V_A) \frac{\mathrm{d}p}{\mathrm{d}T}.$$

Combining this with Eq. [13.4], we obtain the Clausius–Clapeyron equation:

$$\frac{\lambda}{T} = S_B - S_A = (V_B - V_A) \frac{\mathrm{d}p}{\mathrm{d}T}.$$

This equation can also be derived by means of cyclical processes. We carry out a Carnot cycle. (See Fig. 13.2.) We

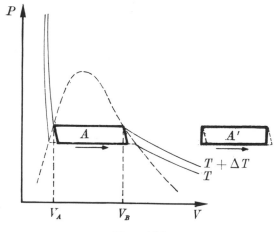

Figure 13.2

have $\lambda/T = W/dT$. Indeed, the work W corresponds to the area A. In the limit as $dT \to 0$, we have $W = dp(V_B - V_A)$, since for small dT the areas A and A' differ from one another only by quantities of higher order. Therefore, we obtain the same result which was obtained above.

Further, if we differentiate

$$\lambda = -T\left[\left(\frac{\partial\Phi}{\partial T}\right)_B - \left(\frac{\partial\Phi}{\partial T}\right)_A\right] = \left[\Phi - T\left(\frac{\partial\Phi}{\partial T}\right)\right]_B - \left[\Phi - T\left(\frac{\partial\Phi}{\partial T}\right)\right]_A$$

with respect to T, we obtain

$$\frac{d\lambda}{dT} = -T\left[\left(\frac{\partial^2\Phi}{\partial T^2}\right)_B - \left(\frac{\partial^2\Phi}{\partial T^2}\right)_A\right]$$
$$+ \frac{dp}{dT}\left\{\left[\left(\frac{\partial\Phi}{\partial p}\right) - T\left(\frac{\partial^2\Phi}{\partial T\,\partial p}\right)\right]_B - \left[\left(\frac{\partial\Phi}{\partial p}\right) - T\left(\frac{\partial^2\Phi}{\partial T\,\partial p}\right)\right]_A\right\},$$

or

$$\frac{d\lambda}{dT} = (c_p)_B - (c_p)_A + \frac{dp}{dT}\left\{\left[V - T\left(\frac{\partial V}{\partial T}\right)_p\right]_B - \left[V - T\left(\frac{\partial V}{\partial T}\right)_p\right]_A\right\}.$$

From the first law we obtained

$$\frac{d\lambda}{dT} = (c_p)_B - (c_p)_A + \frac{dp}{dT}\left[\left\{\left[\left(\frac{\partial E}{\partial V}\right)_T + p\right]\left(\frac{\partial V}{\partial p}\right)_T\right\}_B\right.$$
$$\left. - \left\{\left[\left(\frac{\partial E}{\partial V}\right)_T + p\right]\left(\frac{\partial V}{\partial p}\right)_T\right\}_A\right] + (V_B - V_A)\frac{dp}{dT}.$$

The result from the second law agrees with that from the first law because of the validity of the following relation:

$$\left[\left(\frac{\partial E}{\partial V}\right)_T + p\right]\left(\frac{\partial V}{\partial p}\right)_T = -T\left(\frac{\partial V}{\partial T}\right)_p.$$

This follows from

$$\left[\left(\frac{\partial E}{\partial V}\right)_T + p\right] = T\left(\frac{\partial p}{\partial T}\right)_V$$

with the help of

$$\left(\frac{\partial p}{\partial T}\right)_V = -\frac{(\partial V/\partial T)_p}{(\partial V/\partial p)_T}.$$

The last two relations were derived in Section 12. If we know the thermodynamic function of the liquid or gaseous state of a substance, then we can calculate the equilibrium curve (vapor-pressure curve) for that substance.

f. Unstable states

In Fig. 13.3, both the solid and the dashed curves are isotherms. The dashed arcs represent states (for example,

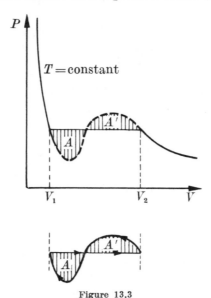

Figure 13.3

superheated liquid or supercooled vapor) which are unstable with respect to arbitrarily small changes of state. If we carry out a cyclical process along these isotherms, then we obtain

$$W = 0 = \oint p \, dV \quad \text{(that is, } A = A')$$

and

$$\int_{V_1}^{V_2} p \, dV = \int_{V_1}^{V_2} p \, dV = p(V_2 - V_1).$$
$$\text{unstable curve} \qquad \text{stable curve}$$

This formula determines the vapor pressure p. Instead of speaking especially of unstable states, we can also say that we know the thermodynamic function [A-10]. The isotherms do not have the form shown for all temperatures T. There exists a critical temperature T_c for which the extreme points of the isotherm coalesce; that is, the isotherm has a saddlepoint. This is defined by

$$\left(\frac{\partial p}{\partial V}\right)_T = 0 \quad \text{and} \quad \left(\frac{\partial^2 p}{\partial V^2}\right)_T = 0 .$$

The critical data of this point are called the *critical temperature* T_c, the *critical pressure* p_c, and the *critical volume* V_c.

14. ENTROPY OF IDEAL GAS MIXTURES

We have the following empirical laws for mixtures of ideal gases:

1. The internal energy of the mixture is equal to the sum of the internal energies of the separate gases:

$$E = \sum_k E_k = \sum_k n_k(c_{V_k} T + E_{0k}) .$$

2. The partial pressure of gas k in the mixture is the pressure which that gas would exhibit if it alone occupied the volume V that the mixture occupies. The pressure p of the mixture is equal to the sum of the partial pressures p_k of the separate gases:

$$p = \sum_k p_k = \frac{RT}{V} \sum_k n_k , \qquad \text{(Dalton's law)}$$

where n_k is the number of moles of gas k in the mixture.

3. If there are two gases occupying volumes V and V', which are separated by a semipermeable membrane (for example, permeable for gas I and impermeable for gas II), then an equilibrium is reached such that $p = p'$. The pressure of I in V is p, and p' is the partial pressure of I

in V'. A sufficient quantity of gas I goes from V to V' for this equilibrium to be achieved.

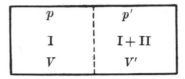

Figure 14.1

With the help of these three empirical laws we can calculate the entropy of mixtures of ideal gases. Two gases can be reversibly mixed in the following way. Let V and V' be two equally large volumes, and let V be inserted into V'. (See Fig. 14.2.) The semipermeable wall A is permeable only to gas I; wall B is permeable only to gas II.

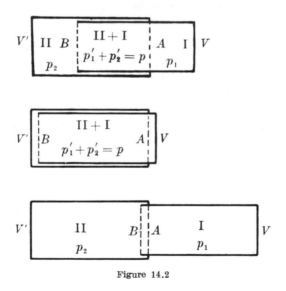

Figure 14.2

Let the movement of V proceed isothermally and infinitely slowly. Both gases are present in the overlapping portions of V and V'. In the two other portions of V and V' are gases I and II separately. If volume V is completely inserted into volume V', then we have a mixture of volume $V = V'$. If we pull the two volumes apart, then the two gases are completely separated. In this way the mixing process is reversible. If we assume that V is moved so slowly that there is always equilibrium, then at every moment, according to 2 and 3, we have $p = p_1' + p_2'$. If V is moved an amount dV, the amount of work done is

$$\delta W = p_1' \, dV + p_2' \, dV - p \, dV = [p_1' - (p - p_2')] \, dV = 0 \,.$$

That is, no work is done either when the gases are mixed or when they are separated. Likewise, there is no change in the internal energy, since the mixing and separating processes are carried out isothermally. Thus, according to the first law, heat is neither added nor removed: $(\delta Q = 0) \to (dS = \delta Q / T = 0)$. Therefore, the entropy has not changed. We now have the following law: The entropy of a gas mixture in volume V at temperature T equals the sum of the entropies of the individual gases when each is in volume V at temperature T. Since the entropy of n_k moles of gas k is

$$S_k = n_k \left(c_{v_k} \log T + R \log \frac{V}{n_k} + a_k \right),$$

we have for the entropy of the mixture

$$S = \sum_k S_k = \sum_k n_k \left(c_{v_k} \log T + R \log \frac{V}{n_k} + a_k \right).$$

Therefore, for mixtures of ideal gases,

$$S = \sum_{k=1}^{m} S_k(n_k, V, T) = S(n_1, n_2, n_3, \dots, n_m, V, T) \,.$$

15. GIBBS'S PARADOX

In two volumes V_1 and V_2, separated by a wall, we have two gases which together have entropy S. If the separating wall is removed, then the entropy changes; let $\bar{S}$ be the entropy of the mixture. We have

$$S = n_1 \left(c_{V_1} \log T + R \log \frac{V_1}{n_1} \right) + n_2 \left(c_{V_2} \log T + R \log \frac{V_2}{n_2} \right),$$

$$\bar{S} = n_1 \left(c_{V_1} \log T + R \log \frac{V}{n_1} \right) + n_2 \left(c_{V_2} \log T + R \log \frac{V}{n_2} \right),$$

$$\bar{S} - S = R \left[n_1 \left(\log \frac{V}{n_1} - \log \frac{V_1}{n_1} \right) + n_2 \left(\log \frac{V}{n_2} - \log \frac{V_2}{n_2} \right) \right]$$

$$= R \left[n_1 \log \frac{V}{V_1} + n_2 \log \frac{V}{V_2} \right] > 0 .$$

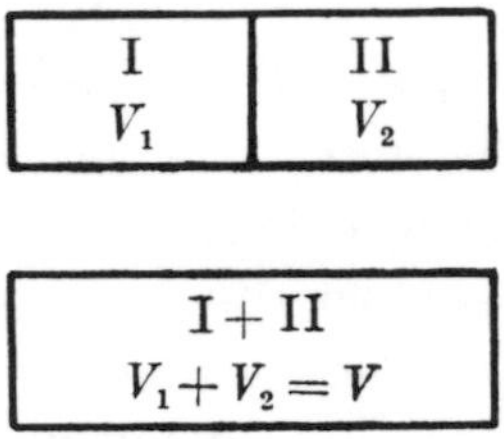

Figure 15.1

The increase in entropy, $\bar{S} - S$, is independent of the nature of the two gases. They must simply be different. If both gases are the same, then the change in entropy is zero; that is,

$$\bar{S} - S = 0 .$$

We see, therefore, that there is no continuous transition between two gases. The increase in entropy is always finite, even if the two gases are only infinitesimally different. However, if the two gases are the same, then the change in entropy is zero. Therefore, it is not allowed to let the difference between two gases gradually vanish. (This is important in quantum theory.)

16. REMARKS ON THE MIXING LAWS

It is not absolutely necessary to use semipermeable walls in order to derive the law concerning gas mixtures. The law can also be derived with the help of force fields. We want to do this with the help of the gravity field. [See E. SCHRÖDINGER, *Z. Physik* **5**, 163 (1921).]

If a gravitational field opposite to the z direction is given, then

$$-\frac{\mathrm{d}p}{\mathrm{d}z} = \varrho g = \frac{M}{v} g = \frac{M}{RT} g p \ .$$

On the assumption that $T(z) = $ constant, this differential equation has the solution

$$p = p_0 \, e^{-\frac{M}{RT} g z} \quad \text{(Barometer formula)} \ ,$$

or

$$\varrho = \varrho_0 e^{-\frac{M}{RT} g z} \ .$$

This effect can be used for separating two gases that have different molar weights. Let there be a heavy gas with

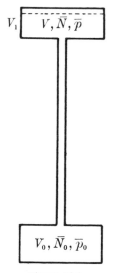

Figure 16.1

molar weight M and density ϱ, and a light gas with molar
weight M' and density ϱ'. We assume the inequality

$$(M - M')gz \gg RT .\tag{16.1}$$

If we have a mixture of both gases in volume V_0, which is
connected with a higher volume V_1 by an ascending tube,
then the light gas rises to V_1 and the heavy gas stays
below. It is assumed that V_0 and V_1 are so chosen that

$$e^{\frac{M}{RT}gz} \gg \frac{V_1}{V_0} \gg e^{\frac{M'}{RT}gz} .\tag{16.2}$$

It is to be noted that V_0 and V_1 can be so chosen that
Eq. [16.2] is satisfied, since Eq. [16.1] implies

$$e^{\frac{M}{RT}gz} \gg e^{\frac{M'}{RT}gz} .$$

Since $\varrho_0/\varrho = e^{Mgz/RT}$ and $\varrho_0'/\varrho' = e^{M'gz/RT}$, we also have that

$$\frac{\varrho_0}{\varrho} \gg \frac{V_1}{V_0} \gg \frac{\varrho_0'}{\varrho'} \quad \text{or} \quad \frac{\varrho_0 V_0}{\varrho V_1} \gg 1 \gg \frac{\varrho_0' V_0}{\varrho' V_1} .$$

That is, the heavy gas stays near the bottom and the
light gas rises. Let us now calculate the work which is
done during the separation. In each intermediate state let
the light gas in the upper region occupy volume V and
have the partial pressure $\bar{p}$. Let the light gas in the lower
region have the partial pressure $\bar{p}_0$. Likewise, let there be $\bar{N}$
moles of the light gas above and $\bar{N}_0$ moles below. We have

$$\bar{N} + \bar{N}_0 = N_0' = \text{constant}, \quad \bar{p} = RT\bar{N}/V, \quad \text{and} \quad \bar{p}_0 = RT\bar{N}_0/V_0;$$

from this it follows that

$$\bar{p}V + \bar{p}_0 V_0 = RT(\bar{N} + \bar{N}_0) = RTN_0' .$$

In the initial state $V = 0$, $\bar{p}_0 = p_0'$, and $\bar{p}V + \bar{p}_0 V_0 = p_0' V_0$,
where p_0' is the partial pressure of the light gas in the

mixture in V_0. In each intermediate state,

$$\overline{p} = \overline{p}_0 \, e^{-\frac{M'}{RT} \, gz},$$

and

$$\overline{p} V + \overline{p}_0 V_0 = \overline{p} \left(V + V_0 e^{+\frac{M'}{RT} \, gz} \right) = p'_0 V_0 \, ;$$

from this, $\overline{p}$ is calculated to be

$$\overline{p} = p'_0 V_0 \Big/ \left(V + V_0 e^{\frac{M'}{RT} \, gz} \right).$$

When the final state, $V = V_1$ and $\overline{p}_0 = 0$, is reached, the amount of work done is

$$W = \int_0^{V_1} \overline{p} \, \mathrm{d}V = p'_0 V_0 \int_0^{V_1} \frac{\mathrm{d}V}{V + V_0 e^{\frac{M'}{RT} \, gz}} = p'_0 V_0 \log \frac{V_1 + V_0 e^{\frac{M'}{RT} \, gz}}{V_0 e^{\frac{M'}{RT} \, gz}}.$$

Because the light gas rises, we have

$$\frac{\varrho'_0 V_0}{\varrho' V_1} = \frac{V_0}{V_1} e^{\frac{M'}{RT} \, gz} \ll 1 \ .$$

Therefore, the work done can be written in good approximation as

$$W = p'_0 V_0 \left(\log \frac{V_1}{V_0} - \frac{M'}{RT} gz \right).$$

If we bring the volume V_1 to the same height as V_0, we thereby gain an amount of work

$$W_{z \to 0} = N' M' gz = p'_0 V_0 \frac{M'}{RT} gz \ .$$

We now change the volume V_1 of the light gas to volume V_0 isothermally; the work done by the gas during that process is

$$W_{V_1 \to V_0} = p'_0 V_0 \log \frac{V_0}{V_1} = - p'_0 V_0 \log \frac{V_1}{V_0}.$$

The total work done is therefore zero. The heat added is

box are the gases (possibly vapors) Cl_2, H_2O, HCl, and O_2, separated from the main box by semipermeable membranes. Membrane A is permeable only to Cl_2, B only to H_2O,

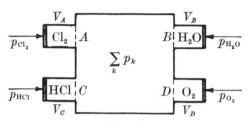

Figure 17.1

C only to HCl, and D only to O_2. Let the pressures in the separate volumes be p_{Cl_2}, p_{H_2O}, p_{HCl}, and p_{O_2}. At equilibrium the pressure in the main box is

$$p = p_{Cl_2} + p_{H_2O} + p_{HCl} + p_{O_2}.$$

Let the entire box be in an infinitely large heat reservoir at temperature T, in order that the temperature remain constant during changes of state.

1. If, as slowly as possible, 2ε moles of Cl_2 are let into volume V_A and 2ε moles of H_2O are let into volume V_B and, at the same time, 4ε moles of HCl are removed from volume V_C and ε moles of O_2 are removed from volume V_D, then, since the number of atoms within the box remains constant, nothing within the box changes. The work done is

$$W_1 = \varepsilon(-2p_{Cl_2} V_{Cl_2} - 2p_{H_2O} V_{H_2O} + 4p_{HCl} V_{HCl} + p_{O_2} V_{O_2}) = \varepsilon RT,$$

where V_{Cl_2} is the volume of one mole of Cl_2 under pressure p_{Cl_2}, etc.

2. We now use a second box in which the reaction reverse to that in 1 occurs, but which works with pressures p'_{Cl_2}, p'_{H_2O}, p'_{HCl}, and p'_{O_2}. Therefore, we bring the two isolated gases which we took from the first box, HCl and O_2, to the

new pressures, p'_{HCl} and p'_{O_2}, by isothermal changes of state. During that process the amount of work done is

$$W_2 = \sum_k \int_{V_k}^{V'_k} n_k p_k \, dV_k = -RT \sum_k \int_{p_k}^{p'_k} n_k \frac{dp_k}{p_k}$$

$$= \varepsilon RT \left(4 \log \frac{p_{HCl}}{p'_{HCl}} + \log \frac{p_{O_2}}{p'_{O_2}} \right).$$

3. We now slowly put 4ε moles of HCl into V_C and ε moles of O_2 into V_D and take 2ε moles of Cl_2 out of V_A and 2ε moles of H_2O out of V_B. During that process the amount of work done is

$$W_3 = -\varepsilon RT \quad \text{(reverse of 1)}.$$

4. We bring the isolated gases which we took from the second box, Cl_2 and H_2O, back to their original pressures by isothermal changes of state. The amount of work done is

$$W_4 = -RT \sum_k \int_{p'_k}^{p_k} n_k \frac{dp_k}{p_k} = \varepsilon RT \left(2 \log \frac{p'_{Cl_2}}{p_{Cl_2}} + 2 \log \frac{p'_{H_2O}}{p_{H_2O}} \right).$$

We have now returned to the initial state. Since there are no changes within the boxes, therefore, according to the second law, the total work done during this cycle must be zero:

$$0 = W_1 + W_2 + W_3 + W_4$$
$$= \varepsilon RT \left(4 \log \frac{p_{HCl}}{p'_{HCl}} + \log \frac{p_{O_2}}{p'_{O_2}} - 2 \log \frac{p_{Cl_2}}{p'_{Cl_2}} - 2 \log \frac{p_{H_2O}}{p'_{H_2O}} \right).$$

From this it follows that

$$4 \log p_{HCl} + \log p_{O_2} - 2 \log p_{Cl_2} - 2 \log p_{H_2O}$$
$$= 4 \log p'_{HCl} + \log p'_{O_2} - 2 \log p'_{Cl_2} - 2 \log p'_{H_2O} = \text{constant}.$$

Since we have carried out the cyclical process at constant temperature, the constant can still be a function of temperature. Therefore, we can make the definition constant $= \log K(T)$.

In general, if M_k is the chemical symbol for gas k, and if ν_k moles of gas k take part in a chemical reaction (where the ν_k are negative quantities if they refer to molecules which are produced), then the reaction equation is

$$\sum_{k=1}^{N} \nu_k M_k = 0 \,,$$

where $N =$ number of gases participating in the reaction. (In our example: $\nu_1 = 2$, $\nu_2 = 2$, $\nu_3 = -4$, $\nu_4 = -1$; $M_1 = Cl_2$, $M_2 = H_2O$, $M_3 = HCl$, $M_4 = O_2$.)

The equilibrium condition is

$$\sum_k \nu_k \log p_k = \log K(T)$$

or

$$\prod_{k=1}^{N} p_k^{\nu_k} = K(T) \,.$$

Since the temperature T and the pressures p_k in the van't Hoff box are constant during the reaction, we can use the Gibbs function,

$$\Phi = E + pV - TS \,,$$

and, since p and T are constant, we have

$$\Delta\Phi = \Delta E + p\Delta V - T\Delta S = 0 \,,$$

where $T\Delta S = \Delta Q = \Delta E + p\Delta V$. This relation is also valid if the mole numbers n_k change, since the equilibrium in the van't Hoff box is not disturbed thereby. Let Φ_k be the Gibbs function of gas k per mole. The change of the mole number n_k because of the chemical reaction is $\Delta n_k = \varepsilon \nu_k$; and since $\Phi = \sum_k n_k \Phi_k$, we have

$$\Delta\Phi = \sum_k \Delta n_k \Phi_k = \varepsilon \sum_k \nu_k \Phi_k = 0 \,.$$

From this we obtain the following equilibrium condition:

$$\sum_k \nu_k \Phi_k = 0 \,.$$

For ideal gases, with

$$\Phi_k = -c_{pk} T \log T + RT \log p_k - i_k RT + E_{0k},$$

we obtain

$$-\left(\sum_k c_{pk} \nu_k\right) T \log T + RT \left(\sum_k \nu_k \log p_k - \sum_k \nu_k i_k\right) + \sum_k \nu_k E_{0k} = 0$$

or

$$\log K(T) = \frac{1}{R} \log T \sum_k c_{pk} \nu_k + \sum_k \nu_k i_k - \frac{1}{RT} \sum_k \nu_k E_{0k}.$$

The heat of reaction at constant pressure, Q_p, is defined by

$$Q_p = T \sum_k \nu_k S_k = -T \sum_k \nu_k \left(\frac{\partial \Phi_k}{\partial T}\right)_{p_k} = -T^2 \sum_k \nu_k \frac{\partial}{\partial T}\left(\frac{\Phi_k}{T}\right).$$

Taking the total derivative with respect to T of the equilibrium condition, $\sum_k \nu_k \Phi_k = 0$, and using the formula

$$\left(\frac{\partial \Phi_k}{\partial p_k}\right)_T \frac{dp_k}{dT} = \frac{\partial}{\partial p_k}(RT \log p_k) \frac{dp_k}{dT} = RT \frac{d}{dT}(\log p_k),$$

we obtain

$$0 = \frac{d}{dT} \sum_k \nu_k \Phi_k = \sum_k \nu_k \left(\frac{\partial \Phi_k}{\partial T}\right)_{p_k} + \sum_k \nu_k \left(\frac{\partial \Phi_k}{\partial p_k}\right)_T \frac{dp_k}{dT},$$

or

$$\frac{Q_p}{RT^2} = \frac{d}{dT}[\log K(T)].$$

We can also derive this formula in the following manner. From the relation

$$\log K(T) = \frac{1}{R} \log T \sum_k \nu_k c_{pk} + \sum_k \nu_k i_k - \frac{1}{RT} \sum_k \nu_k E_{0k},$$

we obtain

$$\frac{d}{dT}[\log K(T)] = \frac{1}{RT} \sum_k \nu_k c_{pk} + \frac{1}{RT^2} \sum_k \nu_k E_{0k}.$$

From the relation

$$Q_p = -T^2 \sum_k \nu_k \frac{\partial}{\partial T}\left(\frac{\Phi_k}{T}\right) = -T^2 \sum_k \nu_k \left[-\frac{c_{pk}}{T} - \frac{E_{0k}}{T^2}\right],$$

we obtain

$$\frac{Q_p}{RT^2} = \frac{1}{RT} \sum_k \nu_k c_{pk} + \frac{1}{RT^2} \sum_k \nu_k E_{0k}. \qquad \text{Q.E.D.}$$

We obtain the same result by means of Carnot cycles. We assume that the second box, with the pressures p_k', has a temperature $T' = T + \Delta T$ different from that of the first box. For ideal gases we have

$$S = -\left(\frac{\partial \Phi}{\partial T}\right)_p = c_p \log T - R \log p + \text{constant}.$$

From this follows the adiabatic equation of state, $S = $ constant:

$$c_p \Delta T - RT \log \frac{p + \Delta p}{p} = 0.$$

In the following, for the sake of simplicity, we neglect quantities of order higher than ΔT. Also, we use the same notation as in the first experiment.

1. We introduce 2ε moles of Cl_2 and 2ε moles of H_2O into the box which is at temperature T, and we remove 4ε moles of HCl and ε moles of O_2. The work done during the process is

$$W_1 = RT\varepsilon \sum_k \nu_k = \varepsilon RT.$$

2. We then bring the gases HCl and O_2 adiabatically and isothermally from (p_k, T) to (p_k', T').
(a) Adiabatically from (p_k, T) to (p_k'', T'):

$$W_a = -\Delta E = -\varepsilon(4c_V^{\text{HCl}} + c_V^{O_2})\Delta T.$$

For each individual gas, we have

$$R \log \frac{p_k''}{p_k} = c_p^k \log \frac{T'}{T} \sim c_p^k \frac{\Delta T}{T}.$$

(b) Isothermally from (p_k'', T') to (p_k', T'):

$$W_b = - RT' \sum_k \int_{p_k''}^{p_k'} n_k \frac{dp_k}{p_k} = - \varepsilon RT' \left(4 \log \frac{p_{HCl}'}{p_{HCl}''} + \log \frac{p_{O_2}'}{p_{O_2}''} \right)$$

$$= - \varepsilon RT' \left(4 \log \frac{p_{HCl}'}{p_{HCl}} + \log \frac{p_{O_2}'}{p_{O_2}} \right)$$

$$+ \varepsilon RT' \left(4 \log \frac{p_{HCl}''}{p_{HCl}} + \log \frac{p_{O_2}''}{p_{O_2}} \right),$$

where the last term is approximately equal to

$$\varepsilon (4 c_p^{HCl} + c_p^{O_2}) \Delta T .$$

Therefore, we find W_2 to be

$$W_2 = W_a + W_b = - \varepsilon RT' \left(4 \log \frac{p_{HCl}'}{p_{HCl}} + \log \frac{p_{O_2}'}{p_{O_2}} \right)$$

$$+ \varepsilon [4 (c_p^{HCl} - c_V^{HCl}) + (c_p^{O_2} - c_V^{O_2})] \Delta T .$$

3. We introduce 4ε moles of HCl and ε moles of O_2 into the box, which is at temperature T', and we remove 2ε moles of Cl_2 and 2ε moles of H_2O. We have $W_3 = - \varepsilon RT'$ (reverse of 1, with T' instead of T).

4. We bring the gases Cl_2 and H_2O adiabatically and iso-thermally from (p_k', T') back to (p_k, T).

(a) Isothermally from (p_k', T') to (p_k'', T'), where p_k'' is so chosen that the adiabatic curve from (p_k, T) passes through (p_k'', T'):

$$W_a = - RT' \sum_k \int_{p_k'}^{p_k''} n_k \frac{dp_k}{p_k} = - \varepsilon RT' \left(-2 \log \frac{p_{Cl_2}'}{p_{Cl_2}''} - 2 \log \frac{p_{H_2O}'}{p_{H_2O}''} \right).$$

(b) Adiabatically from (p_k'', T') to (p_k, T):

$$W_b = - \Delta E = - \varepsilon (2 c_V^{Cl_2} + 2 c_V^{H_2O})(- \Delta T) .$$

For each individual gas we again have

$$R \log \frac{p_k''}{p_k} \sim c_p^k \frac{\Delta T}{T} .$$

Therefore, we find W_4 to be

$$W_4 = -\varepsilon RT' \left(-2 \log \frac{p'_{Cl_2}}{p_{Cl_2}} - 2 \log \frac{p'_{H_2O}}{p_{H_2O}} \right)$$
$$- \varepsilon \left[2(c_p^{Cl_2} - c_V^{Cl_2}) + 2(c_p^{H_2O} - c_V^{H_2O}) \right] \Delta T .$$

The total work done is

$$W = W_1 + W_2 + W_3 + W_4$$
$$= +\varepsilon RT' \left(2 \log \frac{p'_{Cl_2}}{p_{Cl_2}} + 2 \log \frac{p'_{H_2O}}{p_{H_2O}} - 4 \log \frac{p'_{HCl}}{p_{HCl}} - \log \frac{p'_{O_2}}{p_{O_2}} \right)$$
$$= +\varepsilon RT' \sum_k \nu_k \log \frac{p'_k}{p_k} = +\varepsilon RT' [\log K(T') - \log K(T)] .$$

Since we have carried out a Carnot cycle, we have

$$\frac{Q(T')}{T'} = \frac{Q(T)}{T} = \frac{W}{\Delta T} .$$

At constant pressure, $Q(T')$ reduces to the heat of reaction:

$$Q(T') = \varepsilon Q_p \quad (Q_p \text{ per each } \nu_k \text{ moles}) ,$$

$$\frac{Q_p}{T'} = RT' \left[\frac{\log K(T') - \log K(T)}{\Delta T} \right] \sim RT' \frac{d}{dT} [\log K(T)]. \quad \text{Q.E.D.}$$

By using the Carnot cycle, we need not discuss the thermo-dynamic function of the gas mixture in the box. Only properties of ideal gases in relation to semipermeable membranes and gas mixtures are needed.

If c_k is the concentration of gas k in the mixture at pressure p, then

$$p_k = c_k p , \quad \text{where} \quad \sum_k c_k = 1 .$$

We then have

$$\log K(T) = \sum_k \nu_k \log p_k = \sum_k \nu_k \log c_k + \log p \sum_k \nu_k .$$

With the reactions

$$Cl_2 + H_2 \rightleftharpoons 2\,HCl \qquad (\text{with} \quad \nu_1 = 1,\ \nu_2 = 1,\ \nu_3 = -2)$$

and

$$I_2 \rightleftharpoons 2\,I \qquad (\text{with} \quad \nu_1 = 1,\ \nu_2 = -2),$$

there are complications, because the reaction products decompose in the van't Hoff box. In these cases the difficulty is often removed by the introduction of anticatalysts which slow down the reactions. However, this would seem to be arbitrary and unsatisfactory [A-11]. The cyclical processes used by van't Hoff can, by suitable application of external force fields which act differently on the different components, be so generalized that the introduction of unstable states or anticatalysts is superfluous.

Let us consider, for example, the already mentioned dissociation of iodine:

$$I_2 \rightleftharpoons 2\,I\,.$$

In this case it is useful to introduce into the right-hand box a magnetic field of strength H, in which the para-

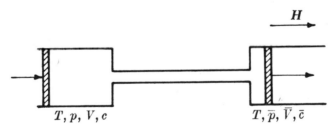

Figure 17.2

magnetic iodine atoms I behave differently from the diamagnetic iodine molecules I_2. Indeed, the former have an additional free energy per gram-atom of

$$F(H) - F(0) = -\tfrac{1}{2}\,\chi H^2,$$

in which χ is generally temperature dependent. This follows from the property of the free energy that

$$\mathrm{d}F = -\, S\,\mathrm{d}T - p\,\mathrm{d}V - M\,\mathrm{d}H$$

when the magnetization is proportional to the field strength,

$$\boldsymbol{M} = \chi(T)\boldsymbol{H}\,.$$

This means that the Gibbs function, $\varPhi = F + pV$, satisfies

$$\mathrm{d}\varPhi = -\, S\,\mathrm{d}T + V\,\mathrm{d}p - M\,\mathrm{d}H$$

in a magnetic field.

Remark: If the sources of the magnetic field are included in the system which is considered, then the work done is given by $\delta W = +H\,\mathrm{d}M$ and the thermodynamic potentials which appear are

$$F' = F + MH \qquad \text{and} \qquad \varPhi' = \varPhi + MH$$

instead of F and $\varPhi$. If the thermodynamic functions F and $\varPhi$ are nevertheless used, then it is unnecessary to consider the sources of the magnetic field. In the following discussion a cyclical process will be considered in which there is a constant field strength (but variable number of free iodine atoms). The various amounts of work $H\Delta M$ which appear in this process cancel one another.

In the left-hand box (zero field strength, pressure p, volume V, temperature T) let there be a total of N gram-atoms of iodine present; of these, Nc are I_2 molecules, and $N(1-2c)$ are free I atoms $(0 < c < \tfrac{1}{2})$. In the right-hand box let the field strength be H and let the temperature T be the same as in the left-hand box; also, let the numbers which correspond to N, p, V, and c be $\bar{N}$, $\bar{p}$, $\bar{V}$, and $\bar{c}$, respectively.

The partial pressures p_1 of the I atoms and p_2 of the I_2

molecules which are predicted by the ideal gas laws are

$$p_1 V = RT \cdot N(1-2c), \qquad p_2 V = RT \cdot Nc,$$

and

$$p V = (p_1 + p_2)V = RT \cdot N(1-c) \; ;$$

and, likewise,

$$\bar{p}_1 \bar{V} = RT \cdot \bar{N}(1-2\bar{c}), \qquad \bar{p}_2 \bar{V} = RT \cdot \bar{N}\bar{c},$$

and

$$\bar{p}\bar{V} = RT \cdot \bar{N}(1-\bar{c}).$$

For the following discussion we need an expression for $V dp$ at constant T. First we have

$$V \, dp = V(dp_1 + dp_2) = NRT\left[(1-2c)\frac{dp_1}{p_1} + c\frac{dp_2}{p_2}\right].$$

Introducing the auxiliary quantities

$$\mu_1 = RT \log p_1 + f_1(T) \quad \text{and} \quad \mu_2 = RT \log p_2 + f_2(T), \quad [17.2]$$

in which the additive functions, which are purely functions of temperature, are at this point arbitrary, we have

$$V \, dp = N(1-2c) \, d\mu_1 + Nc \, d\mu_2 \quad \text{for fixed } T$$

and, likewise,

$$\bar{V} \, d\bar{p} = \bar{N}(1-2\bar{c}) \, d\bar{\mu}_1 + \bar{N}\bar{c} \, d\bar{\mu}_2.$$

We must now draw on the *hydrostatic equilibrium conditions* for the monatomic and diatomic components of the mixture; in the case of ideal gases these components can be considered to be independent of one another. In that part of the tube joining the two vessels in which the field strength H depends on the position coordinate x, there acts on the atoms a force

$$\frac{N(1-2c)}{V} \chi H \frac{dH}{dx}$$

in the x direction; equilibrium is maintained through the gradient $-\mathrm{d}p_1/\mathrm{d}x$ of the partial pressure. Since

$$N(1-2c)/V = p_1/RT,$$

we therefore have

$$\frac{\mathrm{d}p_1}{\mathrm{d}x} - \frac{p_1}{RT}\,\chi H\,\frac{\mathrm{d}H}{\mathrm{d}x} = 0.$$

Multiplying by RT/p_1 and integrating, we obtain

$$RT\log p_1 - \tfrac{1}{2}\,\chi H^2 = \text{constant}$$

or, using the quantity μ_1 introduced in Eq. [17.2],

$$\mu_1 - \tfrac{1}{2}\,\chi H^2 = \text{constant}$$

and therefore

$$\mu_1 = \bar{\mu}_1 - \tfrac{1}{2}\,\chi H^2. \qquad [17.3]$$

Here, the temperature is assumed constant along the tube.

We neglect the diamagnetism of the I_2 molecules, and therefore we can write

$$p_2 = \bar{p}_2, \qquad \mu_2 = \bar{\mu}_2. \qquad [17.4]$$

The expressions for $V\mathrm{d}p$ and $\bar{V}\mathrm{d}\bar{p}$, as well as the hydrostatic equilibrium conditions, Eqs. [17.3] and [17.4], have been extensively generalized for arbitrary substances by Gibbs in his work *On the Equilibrium of Heterogeneous Substances*. (This first appeared during the years 1875-1878. See *Collected Works of J. Willard Gibbs*, Vol. I.) [1] In this connection, see Sections 18 and 19.

We always assume here and in what follows that the existence of a chemical reaction does not influence the general hydrostatic equilibrium conditions in the case that the chemical composition is unchanging.

---

[1] Gibbs first uses the gravitational field as the external force field; he then also considers electric fields.

We are now sufficiently prepared to discuss the chemical equilibrium between I atoms and I_2 molecules in our two boxes. To that end, we assume that at fixed pressures p and $\bar{p}$ and at fixed temperature T, definite concentrations c and $\bar{c}$, and therefore definite partial pressures, will result from the reaction $I_2 \rightleftarrows 2I$; and it is assumed that these concentrations and partial pressures will be such that the hydrostatic equilibrium conditions, Eqs. [17.3] and [17.4], are compatible with chemical equilibrium.

There results a quasi-static (and thus reversible) process if a heat bath is used to maintain the temperature T constant, and if the pistons in the two vessels are infinitely slowly moved in such a way that one of the two volumes V and $\bar{V}$ is made smaller while the other is made larger, such that equilibrium is maintained at every instant. This means that the two pressures p and $\bar{p}$, as well as the concentrations c and $\bar{c}$, remain constant, whereby Eqs. [17.3] and [17.4] are always satisfied. It is convenient (although not necessary) to assume $N=1$, $\bar{N}=0$ in the initial state and $N=0$, $\bar{N}=1$ in the final state (or vice versa), in order that a total of one gram-atom of iodine be brought from the left vessel into the right vessel (or vice versa). For such a reversible process, the potential Φ defined above remains constant and, therefore, the thermodynamic equilibrium condition is

$$\Phi(T, p, c) = \bar{\Phi}(T, \bar{p}, \bar{c}) \qquad \text{(for } N = \bar{N} = 1). \qquad [17.5]$$

If we consider neighboring pressures at constant H and T, then it follows that

$$V \, dp = \bar{V} \, d\bar{p} \, . \qquad [17.6]$$

Remark: This can also be illustrated by means of the following reversible, isothermal cycle.

First, take one gram-atom of iodine from the left vessel into the right vessel at constant pressure (reversible change in concentration). Amount of work $= \bar{p}\bar{V} - pV$.

Without the connecting tube, bring the right-hand vessel to volume $\overline{V} + d\overline{V}$. Amount of work $= \overline{p} d\overline{V}$.

Take the gram-atom of iodine from the right-hand vessel again into the left-hand vessel.

Amount of work $= - (\overline{p} + d\overline{p})(\overline{V} + d\overline{V}) + (p + dp)(V + dV)$.

Finally, without the connecting tube, take the left vessel from $V + dV$ to V. Amount of work $= - p dV$.

The net work is $V dp - \overline{V} d\overline{p}$ and, according to the second law, it must vanish.

From the general expressions for $V dp$ and $\overline{V} d\overline{p}$ (with $N = \overline{N} = 1$), we have

$$(1 - 2c) d\mu_1 + c d\mu_2 = (1 - 2\overline{c}) d\overline{\mu}_1 + \overline{c} d\overline{\mu}_2 ,$$

whereas the hydrostatic equilibrium conditions imply

$$d\overline{\mu}_1 = d\mu_1 , \qquad d\overline{\mu}_2 = d\mu_2 .$$

Therefore, we obtain

$$(c - \overline{c})(2 d\mu_1 - d\mu_2) = 0 .$$

Because $c \neq \overline{c}$ (from the hydrostatic equilibrium condition it follows that $c < \overline{c}$), therefore

$$\frac{d\mu_2}{d\mu_1} = 2 \quad \text{and} \quad \mu_2 - 2\mu_1 = \text{function of temperature only} .$$

We now fix the functions f_1 and f_2, which occur in the definition of μ_1 and μ_2, in such a way that

$$\mu_2 - 2\mu_1 = 0 . \tag{17.7}$$

At constant T,

$$d\Phi = V dp = d\mu_1 = \tfrac{1}{2} d\mu_2 ,$$

and we make the definitions of f_1 and f_2 unique by requiring

$$\Phi = \mu_1 = \tfrac{1}{2} \mu_2 \tag{17.8}$$

or

$$\Phi - RT \log p_1 + f_1(T) = \tfrac{1}{2}[RT \log p_2 + f_2(T)]. \quad [17.9]$$

Using Eq. [17.4], we then have

$$\bar{\Phi} = \Phi = RT \log \bar{p}_1 + f_1(T) - \tfrac{1}{2} \chi H^2 = \tfrac{1}{2}[RT \log \bar{p}_2 + f_2(T)].$$

The equilibrium constant K was defined by

$$\log K = 2 \log p_1 - \log p_2, \qquad K = \frac{p_1^2}{p_2}. \quad [17.10]$$

Using Eq. [17.9], this can be written as

$$\log K = \frac{f_2 - 2f_1}{RT}, \qquad\qquad [17.11]$$

which is a function of temperature alone. Correspondingly,

$$\bar{K} = \frac{\bar{p}_1^2}{\bar{p}_2} \quad \text{and} \quad \log \bar{K} = \log K + 2 \frac{\tfrac{1}{2}\chi H^2}{RT}.$$

Now we apply the general thermodynamic formula,

$$E + pV = - T^2 \frac{\partial}{\partial T}\left(\frac{\Phi}{T}\right) = \Phi - T \frac{\partial \Phi}{\partial T},$$

in which the pressure is to be held constant during differentiation with respect to T. This expression, combined with Eq. [17.11], only gives the concentration c implicitly as a function of p and T; the same is true for the partial pressures

$$p_1 = p \frac{1 - 2c}{1 - c} \quad \text{and} \quad p_2 = p \frac{c}{1 - c}.$$

If, however, before differentiation with respect to T, we make use of Eq. [17.9] and write

$$\Phi = (1 - 2c)\left[RT \log\left(p \frac{1 - 2c}{1 - c}\right) + f_1(T)\right]$$
$$+ c\left[RT \log\left(p \frac{c}{1 - c}\right) + f_2(T)\right],$$

then, as a result of Eq. [17.11], we formally have

$$\left(\frac{\partial \Phi}{\partial c}\right)_{p,T} = 0 \, ,$$

and it follows that

$$E + pV = (1 - 2c)(- T^2) \frac{\mathrm{d}}{\mathrm{d}T}\left(\frac{f_1}{T}\right) + c(- T^2) \frac{\mathrm{d}}{\mathrm{d}T}\left(\frac{f_2}{T}\right).$$

As long as the coefficients of $(1 - 2c)$ and c in this expression are independent of p, the positive quantity

$$Q_p = T^2 \frac{\mathrm{d}}{\mathrm{d}T} \frac{f_2 - 2f_1}{T} \qquad [17.12]$$

can be considered as the heat which must be added in order to dissociate one gram-molecule of I_2 into two gram-atoms of I at constant pressure. Correspondingly, in the right-hand box (field strength H) we have

$$\bar{E} + \bar{p}\bar{V} = (1 - 2\bar{c})\left[- T^2 \frac{\mathrm{d}}{\mathrm{d}T}\left(\frac{f_1}{T}\right) + T^2 \frac{1}{2} H^2 \frac{\mathrm{d}}{\mathrm{d}T}\left(\frac{\chi}{T}\right)\right]$$

$$+ \bar{c}(- T^2) \frac{\mathrm{d}}{\mathrm{d}T}\left(\frac{f_2}{T}\right);$$

therefore,

$$\bar{Q}_p = Q_p + 2 \cdot \frac{1}{2} H^2 T^2 \frac{\mathrm{d}}{\mathrm{d}T}\left(\frac{\chi}{T}\right). \qquad [17.13]$$

(For χ in this expression we can use Curie's law, $\chi = C/T$; therefore, $\bar{Q}_p < Q_p$.)

(For temperature-independent specific heats at constant pressure we have

$$- T^2 \frac{\mathrm{d}}{\mathrm{d}T}\left(\frac{f_1}{T}\right) = c_{p1} T + E_{01}$$

and

$$- T^2 \frac{\mathrm{d}}{\mathrm{d}T}\left(\frac{f_2}{T}\right) = c_{p2} T + E_{02} \, ;$$

therefore,

$$f_1 = - c_{p1} T \log T + E_{01} - i_1 RT$$

and

$$f_2 = - c_{p2} T \log T + E_{02} - i_2 RT ,$$

as we know.)

Furthermore, from Eqs. [17.11] and [17.12] follow the familiar equations

$$RT^2 \frac{d}{dT} \log K = Q_p \quad \text{and} \quad RT^2 \frac{d}{dT} \log \bar{K} = \bar{Q}_p . \qquad [17.14]$$

Thus, the reversible process considered here, together with the hydrostatic equilibrium conditions in an external field, is sufficient for determining the thermodynamic properties of the partially dissociated mixture of I_2 and I. (See below for the generalization to nonideal gases.)

I would like to add some short remarks concerning the reaction

$$H_2 + Cl_2 \rightleftharpoons 2 \, HCl ,$$

in which the numbers

$$A_{Cl} = 2N_{Cl_2} + N_{HCl}$$

and

$$A_H = 2N_{H_2} + N_{HCl}$$

remain constant. The concentrations of the participating gases are determined uniquely by chemical equilibrium only when, in addition to p and T, the quotient A_{Cl}/A_H is given. Therefore, in order that the chemical composition within the box under consideration remain constant, a further condition must be fulfilled. However, the isolated component gases, Cl_2 and H_2, are both stable, and we can imagine introducing semipermeable partitions for these gases. As a matter of fact, it will suffice to introduce such a partition for only one of these gases; let us say H_2, for example.

Thus, we first consider a box which consists of two subsystems, a and b, which are separated by a partition which is semipermeable to H_2. There is only H_2 in system a, whereas system b contains an equilibrium mixture of H_2, Cl_2, and HCl. Each of the two systems a and b is equipped with a movable piston. However, the hydrostatic equilibrium condition,

$$\mu_{H_2}^a = \mu_{H_2}^b , \qquad [17.15]$$

is always to be satisfied, so that we can drop the indices a and b from μ_{H_2}. Consider fixed T and fixed total amounts A_{Cl} and A_H of Cl and H. Then, using Eq. [17.15], we obtain

$$\begin{aligned}
d\Phi &= V_a\,dp_a + V_b\,dp_b \\
&= N_{H_2}\,d\mu_{H_2} + N_{Cl_2}\,d\mu_{Cl_2} + N_{HCl}\,d\mu_{HCl}
\end{aligned}$$

for the Gibbs function Φ of the entire box (sum of the contributions from a and b). Here N_{H_2} is the total molar number for H_2 in a and b. (The two parts of N_{H_2} do not appear separately in the following discussion.)

A second box, consisting of two parts of the same sort as the parts of the first box, is to be situated in external force fields. In these force fields there are differences in the potential energy per mole for the three gases H_2, Cl_2, and HCl. We label these potential energies by $E_{pot.k}$, where the index k can take on three values. For $k = H_2$ we assume that the value of $E_{pot.H_2}$ is the same in both parts, a and b, of the box.

If part b of the left-hand box is connected with part b of the right-hand box by means of a tube, then the hydrostatic equilibrium conditions are

$$\bar{\mu}_k + E_{pot.k} = \mu_k , \qquad [17.16]$$

where again a quantity which refers to the right-hand box is indicated by a bar over the appropriate symbol.

We can imagine that the two pistons of the left-hand box and those of the right-hand box are moved in such a way that all pressures and concentrations in each part of the boxes remain constant during the irreversible process that occurs. (The temperature is held constant by means of a heat reservoir; the sums $A_H + \bar{A}_H$ and $A_{Cl} + \bar{A}_{Cl}$ are trivially constant during the process.) Therefore, the sum

Figure 17.3

of the functions Φ and $\bar{\Phi}$ of the left-hand and right-hand boxes must remain constant during this process. In what follows we always indicate the values of a quantity in the final state by a prime and the values in the initial state by no prime. With this notation we have

$$\Phi + \bar{\Phi} = \Phi' + \bar{\Phi}' . \qquad [17.17]$$

Now, let us consider the same process at neighboring pressures $p_a + \mathrm{d}p_a$, $p_b + \mathrm{d}p_b$, $\bar{p}_a + \mathrm{d}\bar{p}_a$, and $\bar{p}_b + \mathrm{d}\bar{p}_b$; however, let us keep the same values of A_H, A_{Cl}, $\bar{A}_H$, and $\bar{A}_{Cl}$ in the initial state and of A'_H, A'_{Cl}, $\bar{A}'_H$, and $\bar{A}'_{Cl}$ in the final state. Exactly as before, the two processes at constant pressures and concentrations can be thought of as parts of a single cyclical process; this is accomplished by introducing two subsidiary processes in which the left-hand and right-hand boxes are uncoupled.

Because the hydrostatic equilibrium condition implies $\mathrm{d}\mu_k = \mathrm{d}\bar{\mu}_k$, we obtain

$$\sum_k (N_k + \bar{N}_k)\,\mathrm{d}\mu_k = \sum_k (N'_k + \bar{N}'_k)\ \mathrm{d}\mu_k \qquad [17.18]$$

from Eqs. [17.16] and [17.17].

However, as a result of the reaction we have, with a

definite $\lambda \neq 0$,

$$N'_{H_2} + \bar{N}'_{H_2} - (N_{H_2} + \bar{N}_{H_2}) = \lambda \,,$$

$$N'_{Cl_2} + \bar{N}'_{Cl_2} - (N_{Cl_2} + \bar{N}_{Cl_2}) = \lambda \,,$$

and

$$N'_{HCl} + \bar{N}'_{HCl} - (N_{HCl} + \bar{N}_{HCl}) = -2\lambda \,.$$

(This only expresses the conservation of $A_{Cl} + \bar{A}_{Cl}$ and $A_H + \bar{A}_H$ during the process.)

Therefore, from Eq. [17.18] we have

$$d\mu_{H_2} + d\mu_{Cl_2} - 2 \, d\mu_{HCl} = 0 \,.$$

We choose the additive functions of temperature which are contained in the μ_k in such a way that this relation holds for the μ_k themselves:

$$\mu_{H_2} + \mu_{Cl_2} - 2\mu_{HCl} = 0 \,. \qquad [17.19]$$

Furthermore, from this condition it follows that

$$d\Phi = \tfrac{1}{2} (A_H \, d\mu_{H_2} + A_{Cl} \, d\mu_{Cl_2})$$

for the entire left-hand box at constant A_H, A_{Cl}, and T. (Similarly, for the right-hand box.)

The additive functions of temperature in the μ_k are then completely determined by the additional condition

$$\Phi = \tfrac{1}{2} (A_H \mu_{H_2} + A_{Cl} \mu_{Cl_2}) \,.$$

Here we have already written all the relations in such a way that they are correct for arbitrary substances (see Section 19). It is only the concept of partial pressure p_k and the special form of μ_k,

$$\mu_k = RT \log p_k + f_k(T) \,,$$

which are peculiar to ideal gases.

18. GIBBS'S VARIATIONAL METHOD

Gibbs makes use of virtual changes of state of closed systems. He assumes that the thermodynamic function exists for the states in the neighborhood of stable states. For the change in entropy we must have

$$(\Delta S)_{E,V} = S_{\text{neighboring state}} - S_{\text{stable state}} \leqslant 0 .$$

If it were that $(\Delta S)_{E,V} \geqslant 0$, then, since the entropy is an increasing function (second law), the change in entropy could occur spontaneously; that is, the neighboring state would be stable. Thus, we have the following law: A closed system at fixed volume and internal energy is in equilibrium when its entropy has the largest value consistent with the given volume and internal energy.

If T and V, or T and p, are held constant in a system instead of E and V, then we have the inequality

$$(\Delta F)_{T,V} \geqslant 0 \quad \text{or} \quad (\Delta \Phi)_{T,p} \geqslant 0 , \quad \text{respectively} .$$

Therefore, a closed system at fixed volume that is held at a given temperature is in equilibrium when its free energy has the smallest value consistent with this volume and temperature. Also, a closed system at fixed pressure and temperature is in equilibrium when the Gibbs function has the smallest value consistent with these conditions. For reversible changes of state, the equilibrium conditions are

$$(\delta S)_{E,V} = 0 , \quad (\delta F)_{V,T} = 0 , \quad (\delta \Phi)_{T,p} = 0 .$$

Whether these states are stable is determined by the second variation. We must have

$$(\delta^2 S)_{E,V} < 0 , \quad (\delta^2 F)_{V,T} > 0 , \quad (\delta^2 \Phi)_{T,p} > 0 .$$

That is, the second variations must be definite quadratic forms with the correct sign.

19. APPLICATIONS (FIRST VARIATION)

a. Arbitrary substances

Let the Gibbs function for an arbitrary substance be given. If in the substance there are N_k moles of a basic substance k, then

$$\Phi = \Phi(T, p, N_1, N_2, ..., N_m).$$

The possible variations at constant pressure and temperature are

$$\nu_k \varepsilon = \delta N_k,$$

corresponding to a reaction assumed to be possible. The first variation of Φ is

$$\delta\Phi = \sum_k \left(\frac{\partial\Phi}{\partial N_k}\right)_{T,p} \delta N_k = \varepsilon \sum_k \nu_k \mu_k,$$

where

$$\mu_k = (\partial\Phi/\partial N_k)_{T,p}$$

is called the *chemical potential*. Because $(\delta\Phi)_{T,p} = (\delta F)_{T,V}$ follows immediately from $\Phi(T, p) = F(T, V) + pV$, we also have

$$\mu_k = \left(\frac{\partial\Phi}{\partial N_k}\right)_{T,p} = \left(\frac{\partial F}{\partial N_k}\right)_{T,V}.$$

Using $\sum_k \nu_k \Phi_k = 0$, we thereby obtain the general equilibrium condition

$$\sum_k \nu_k \mu_k = 0.$$

b. Homogeneity properties

The free energy and the Gibbs function are homogeneous functions of the first degree in the variables V, N_k, and in the variables N_k, respectively:

$$F(T, \lambda V, \lambda N_k) = \lambda F(T, V, N_k),$$
$$\Phi(T, p, \lambda N_k) = \lambda \Phi(T, p, N_k).$$

From this, if the temperature is held constant, follows the Euler relation for F:

$$F = V\left(\frac{\partial F}{\partial V}\right)_{T,N_k} + \sum_k \left(\frac{\partial F}{\partial N_k}\right)_{T,V} N_k.$$

With

$$\mathrm{d}F = -p\,\mathrm{d}V + \sum_k \mu_k\,\mathrm{d}N_k\,,$$

we have

$$\left(\frac{\partial F}{\partial V}\right)_{T,N_k} = -p \quad \text{and} \quad \left(\frac{\partial F}{\partial N_k}\right)_{T,V} = \mu_k\,,$$

and therefore

$$F = \sum_k \mu_k N_k - pV\,.$$

This allows us to set

$$F(T, V, N_k) = Vf(T, n_k)\,,$$

where $n_k = N_k/V = \varrho_k/M_k$. ($M_k = $ molecular weight of substance k.) For the chemical potential we then obtain

$$\mu_k = \left(\frac{\partial F}{\partial N_k}\right)_{T,V} = \left(\frac{\partial f}{\partial n_k}\right)_T\,.$$

At constant T and p, the Euler relation for Φ is

$$\Phi = \sum_k N_k \left(\frac{\partial \Phi}{\partial N_k}\right)_{p,T}$$

or

$$\Phi = \sum_k \mu_k N_k\,.$$

On the one hand,

$$\mathrm{d}\Phi = \mathrm{d}(F + pV) = V\,\mathrm{d}p + \sum_k \mu_k\,\mathrm{d}N_k\,,$$

and on the other hand,

$$\mathrm{d}\Phi = \sum_k \mu_k\,\mathrm{d}N_k + N_k\,\mathrm{d}\mu_k\,.$$

We therefore have the important relation (Gibbs):

$$V\,\mathrm{d}p = \sum_k N_k\,\mathrm{d}\mu_k \quad \text{(for fixed } T)\,.$$

c. *Ideal gases*

Let $N = \sum_k N_k$ be the total molar number of a mixture of ideal gases. Writing the partial pressure as

$$p_k = \frac{N_k}{N} p ,$$

we obtain

$$\Phi = \sum_k N_k \left(RT \log p \frac{N_k}{N} - c_{pk} T \log T - i_k RT + E_{0k} \right)$$

for the Gibbs function of the mixture. Making use of the relation

$$\frac{\partial}{\partial N_k} \left[\sum_i N_i \log \frac{N_i}{N} \right] = \frac{\partial}{\partial N_k} \left[\sum_i N_i \log N_i - N \log N \right]$$
$$= \log N_k - \log N ,$$

we obtain

$$\mu_k = \left(\frac{\partial \Phi}{\partial N_k} \right) = - c_{pk} T \log T + RT \log p \frac{N_k}{N} - i_k RT + E_{0k}$$

for the chemical potential. Therefore,

$$\mu_k = RT \log \frac{N_k}{N} p + f_k(T) ;$$

or, with $n_k = N_k / V \sim N_k p / N$,

$$\mu_k = RT \log n_k + \varphi_k(T) .$$

Both $f_k(T)$ and $\varphi_k(T)$ are functions of temperature only.

d. Semipermeable walls

Consider the case of a semipermeable wall. In volume V' let there be N_1' moles of substance 1, and in volume V'' let there be N_k'' moles of substance k $(k = 1, 2, 3, ..., m)$. Let the wall between V' and V'' be permeable to substance 1.

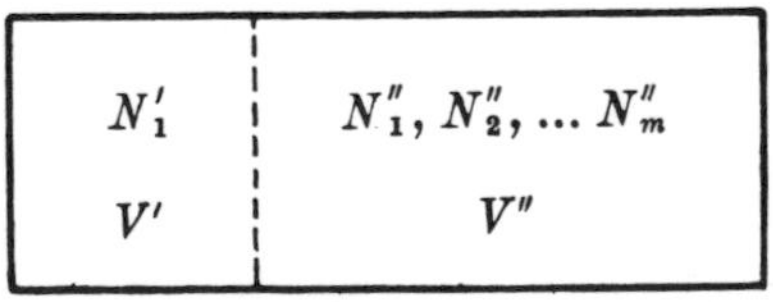

Figure 19.1

The entire system is to be in an infinitely large heat reservoir at temperature T. Since T and V are constant we use the free energy $F(T, V, N_k)$. The possible variations at constant volume and temperature are

$$\delta F = \mu_1(n_1', 0, 0, \ldots, 0)\, \delta N_1' + \mu_1(n_1'', n_2'', n_3'', \ldots, n_m'')\, \delta N_1'' = 0 \,.$$

Since the molecules leaving V' go to V'', we must have

$$\delta N_1' + \delta N_1'' = 0 \,.$$

From these two relations we obtain the condition

$$\mu_1(n_1', 0, 0, \ldots, 0) = \mu_1(n_1'', n_2'', n_3'', \ldots, n_m'') \,.$$

The concept of partial pressure does not appear here. In this example the possible variation is an actual reaction. For ideal gases we obtain

$$\log n_1' = \log n_1'' \quad \text{or} \quad n_1' = n_1'';$$

that is,

$$N_1'/V' = N_1''/V'' \,.$$

If we define $p_1' = RTN_1'/V'$ and $p_1'' = RTN_1''/V''$, which actually correspond to the partial pressures, we obtain

$$p' = p'' \,.$$

e. Force fields

Consider now a mixture of arbitrary substances in a force field, and let $f(n_1, n_2, n_3, \ldots, n_m, T)$ be the free energy per unit volume. Let the potential energy $E_{k,\,\text{pot}}$ of a constituent substance be proportional to the mole number N_k. The total free energy is

$$F = \int_V \left[f(n_k, T) + \sum_k n_k E_{k,\,\text{pot}}(x) \right] dV \,,$$

where $n_k(x)$ depends on position. With respect to variation of the $n_k(x)$, F is to have a minimum value; thus,

$$\delta F = \int_V \left[\left(\frac{\partial f}{\partial n_k} \right)_T + E_{k,\,\text{pot}}(x) \right] \delta n_k \, dV = 0 \,.$$

Since the substance can only be displaced in the force field, none of it is lost. Therefore,

$$\int_V \delta n_k \, \mathrm{d}V = 0 \, .$$

From the above two relations follows

$$\left(\frac{\partial f}{\partial n_k}\right)_T + E_{k,\,\text{pot}}(\boldsymbol{x}) = \text{constant} = a_k$$

or

$$\mu_k(n_1, n_2, n_3, \ldots, n_m) + E_{k,\,\text{pot}}(\boldsymbol{x}) = a_k \, .$$

This is the generalized barometer formula which was originally derived by Gibbs for the gravitational field.

f. Osmotic pressure

The Gibbs variational method can be quite generally applied to solutions.[2] Denote the solvent by 1, and the solute by 2. In V_0 let there be N_1^0 moles of solvent, and in V

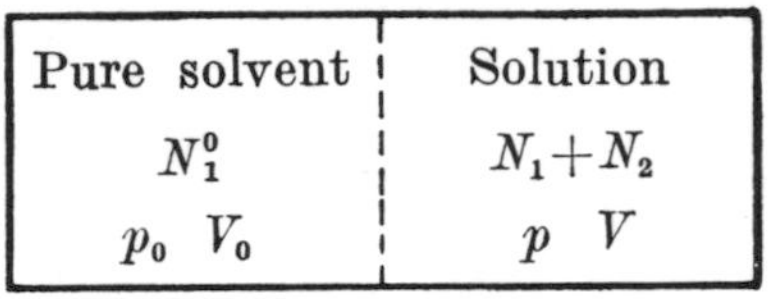

Pure solvent	Solution
N_1^0	$N_1 + N_2$
$p_0 \quad V_0$	$p \quad V$

Figure 19.2

let there be a solution consisting of N_1 moles of solvent and N_2 moles of solute, separated from V_0 by a semi-permeable membrane which is permeable only to the solvent. The concentration in V is then

$$c = N_2/N_1 \qquad \text{(solute divided by solvent)} \, .$$

With semipermeable membranes the equilibrium condition is

$$\mu_1(p_0, N_1^0, 0) = \mu_1(p, N_1, N_2) \, .$$

[2] References for Sections *f* and *g*: J. W. GIBBS, *Nature* **55**, 461 (1897); also, "Semipermeable Films and Osmotic Pressure," *Collected Works of J. Willard Gibbs*, Vol. I, p. 413.

Since the Gibbs function is a homogeneous function of the first degree, we can write

$$\Phi(T, p, N_1, N_2) = N_1 f(T, p, c) \qquad \text{and} \qquad \Phi_1^0 = N_1^0 f(T, p, 0) \,.$$

Therefore,

$$\mu_1 = \left(\frac{\partial \Phi}{\partial N_1}\right)_{N_2} = f - c\left(\frac{\partial f}{\partial c}\right) \quad \text{and} \quad \mu_2 = \left(\frac{\partial \Phi}{\partial N_2}\right)_{N_1} = \left(\frac{\partial f}{\partial c}\right).$$

If π is the osmotic pressure, then we have the equilibrium condition

$$\mu_1(p_0 + \pi, c) = \mu_1(p_0, 0) \,, \quad \text{where} \quad p = p_0 + \pi \,. \quad \cdot$$

This relation can be transformed into

$$\mu_1(p_0, 0) - \mu_1(p_0 + \pi, 0) = \mu_1(p_0 + \pi, c) - \mu_1(p_0 + \pi, 0) \,.$$

Because $(\partial \Phi/\partial p)_T = V$ and $(\partial \Phi/\partial N_1)_{T,p} = \mu_1$, it follows that $(\partial \mu_1/\partial p)_T = (\partial V/\partial N_1)_{T,p}$. For $c = 0$, that is, if V contains only solvent, then

$$\left(\frac{\partial \mu_1}{\partial p}\right)_T = \frac{V}{N_1} \,.$$

Therefore,

$$\mu_1(p_0 + \pi, 0) - \mu_1(p_0 + \pi, c)$$
$$= -\mu_1(p_0, 0) \mid \mu_1(p_0 \mid \pi, 0) = \int_{p_0}^{p_0+\pi} \frac{V}{N_1} \, \mathrm{d}p \,.$$

If we neglect the compressibility of the solvent, we obtain

$$\mu_1(p_0 + \pi, 0) - \mu_1(p_0 + \pi, c) \sim \left(\frac{V}{N_1}\right)\pi = \left(\frac{M_1}{\varrho_1}\right)\pi \,.$$

The dependence of the chemical potential on the concentration c cannot be determined from thermodynamics.

g. Equilibrium between the solution and the vapor of the solute

If the solute is volatile, then there is equilibrium between the vapor of the solute and the solution. If $\bar{p}$ is the vapor pressure, then

$$\mu_2(p, c) = \mu_2(\overline{p}) = \left(\frac{\partial f}{\partial c}\right).$$

Because of the ideal gas laws, we have

$$\mu_2(\overline{p}) = RT \log \overline{p} + f(T) \qquad \text{(chemical potential of gases)}.$$

Therefore,

$$\left(\frac{\partial f}{\partial c}\right) = RT \log \overline{p} + f(T).$$

Using Henry's empirical law,

$$\overline{p} = \alpha c(1 + \varkappa c + ...),$$

we obtain

$$\left(\frac{\partial f}{\partial c}\right) = A_1 + RT \log c + RT \log (1 + \varkappa c + ...).$$

Integration gives

$$f = A_0 + A_1 c + RT(c \log c - c) + RT \left(\frac{\varkappa}{2} c^2 + ...\right);$$

from this it follows that

$$\mu_1 = f - c\left(\frac{\partial f}{\partial c}\right) = A_0 - RTc - RT \frac{\varkappa}{2} c^2 - ...$$
$$= A_0 - RTc \left(1 + \frac{\varkappa}{2} c + ...\right),$$

where $A_0, A_1, ..., \varkappa$ can be functions of p and T. We have

$$\mu_1(p_0 + \pi, 0) - \mu_1(p_0 + \pi, c)$$
$$= \frac{V}{N_1} \pi = \frac{M_1}{\varrho_1} \pi = RTc \left(1 + \frac{\varkappa}{2} c + ...\right),$$

or

$$\pi = \frac{RTcN_1}{V} \left(1 + \frac{\varkappa}{2} c + ...\right).$$

Here $RTcN_1/V = RTN_2/V$ is the pressure of an ideal gas whose volume is the same as that of the solution and which has the same number of molecules as the solute.

h. Equilibrium between the solution and the vapor of the solvent

Of course, there is also an equilibrium between the vapor of the solvent and the solution. Let $\hat{p}$ be the vapor pressure of the solvent. We have

$$\mu_1(p, c) = \mu_{1v}(\hat{p}).$$

Between the pure solvent and its vapor there is the relation

$$\mu_1(p, 0) = \mu_{1v}(p).$$

Therefore,

$$\mu_1(p, c) - \mu_1(p, 0) = \mu_{1v}(\hat{p}) - \mu_{1v}(p) = \int_p^{\hat{p}} \left(\frac{V}{N_1}\right) dp = RT \log \frac{\hat{p}}{p}.$$

According to the definition of osmotic pressure,

$$\mu_1(p, c) = \mu_1(p - \pi, 0).$$

Therefore,

$$\mu_1(p, c) - \mu_1(p, 0) = \mu_1(p - \pi, 0) - \mu_1(p, 0) = RT \log \frac{\hat{p}}{p}.$$

Neglecting the compressibility of the solution, we have

$$\mu_1(p - \pi, 0) - \mu_1(p, 0) = \int_p^{p-\pi} \left(\frac{V}{N_1}\right) dp = -\left(\frac{V}{N_1}\right)\pi = RT \log \frac{\hat{p}}{p},$$

or

$$\left(\frac{V}{N_1}\right)\pi = RT \log \frac{p}{\hat{p}};$$

that is,

$$p > \hat{p}.$$

The above formula gives the relation between osmotic pressure and the change in vapor pressure which comes about because there is a solution instead of a pure solvent. From this formula, changes in boiling point and freezing point can be calculated as a function of osmotic pressure.

i. Vapor pressure above a spherical surface

We consider a sphere of radius r which is to contain n moles of fluid per cm³. At constant pressure we assume that n is constant. Then

$$A = 4\pi r^2 , \qquad \mathrm{d}A = 8\pi r\,\mathrm{d}r ,$$

$$N = n\,\frac{4\pi}{3}\,r^3 , \quad \mathrm{d}N = n4\pi r^2\,\mathrm{d}r , \quad \mathrm{d}A = \frac{2}{nr}\,\mathrm{d}N .$$

Let γ be the surface tension. Then

$$\delta W = p\,\mathrm{d}V - \gamma\,\mathrm{d}A .$$

With $\bar{\Phi} = \Phi + \gamma A$, we obtain

$$\mathrm{d}\bar{\Phi} = - S\,\mathrm{d}T + V\,\mathrm{d}p + \mu\,\mathrm{d}N + \gamma\,\mathrm{d}A ,$$

$$(\delta\bar{\Phi}_l)_{T,p} = \mu_l(p)\,\delta N_l + \gamma\,\mathrm{d}A = \left[\mu_l(p) + \gamma\,\frac{2}{nr}\right]\delta N_l ,$$

$$(\delta\bar{\Phi}_v)_{T,p} = \mu_v(p)\,\delta N_v .$$

With $\delta(N_l + N_v) = 0$, we have

$$\delta(\bar{\Phi}_l + \bar{\Phi}_v)_{T,p} = \left[\mu_l(p) + \gamma\,\frac{2}{nr} - \mu_v(p)\right]\delta N_l = 0$$

and

$$\mu_l(p') + \gamma\,\frac{2}{nr} = \mu_v(p') .$$

Here p' is to denote the vapor pressure above the spherical surface. This formula can also be derived from

$$\delta(F_l + F_v)_{T,V} = 0 \quad \text{with} \quad \delta(V_l + V_v) = 0 .$$

Taking the limit $r \to \infty$, we obtain, for a plane surface (vapor pressure p),

$$\mu_l(p) = \mu_v(p) .$$

Thus,

$$\gamma \frac{2}{nr} = [\mu_v(p') - \mu_l(p')] - [\mu_v(p) - \mu_l(p)] ;$$

or, in terms of the molar volume v,

$$\gamma \frac{2}{nr} = \int_p^{p'} (v_v - v_l)\, dp .$$

This equation is exact.

Furthermore, for an ideal gas and an incompressible fluid,

$$\gamma \frac{2}{nr} = RT \log \frac{p'}{p} - v_l(p' - p) .$$

Thus, the vapor pressure p' depends on the radius r. The smaller r is, the larger p' is at constant p (because the derivative with respect to p' of the right-hand side of the formula is $RT/p' - v_l = v_v - v_l > 0$).

20. COMMENTS ON THE SECOND VARIATION

Let a homogeneous substance in a volume V be separated into two fixed parts of volumes V_1 and V_2 by an

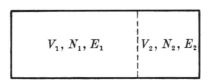

Figure 20.1

imaginary partition. We want to find the conditions for stable equilibrium. We have

$$V_1 + V_2 = V, \qquad N_1 + N_2 = N,$$

$$E_1 + E_2 = E, \quad \text{and} \quad S_1 + S_2 = S.$$

The first variation is

$$\delta S = \left(\frac{\partial S_1}{\partial E_1}\right) dE_1 + \left(\frac{\partial S_2}{\partial E_2}\right) dE_2 + \left(\frac{\partial S_1}{\partial N_1}\right) dN_1 + \left(\frac{\partial S_2}{\partial N_2}\right) dN_2 = 0 ,$$

with the subsidiary conditions

$$dE_1 + dE_2 = 0 \quad \text{and} \quad dN_1 + dN_2 = 0 .$$

Therefore, the equilibrium conditions are

$$\left(\frac{\partial S_1}{\partial E_1}\right) = \left(\frac{\partial S_2}{\partial E_2}\right) \qquad [20.1]$$

and

$$\left(\frac{\partial S_1}{\partial N_1}\right) = \left(\frac{\partial S_2}{\partial N_2}\right) . \qquad [20.2]$$

From Eq. [20.1] follows $T_1 = T_2$.

Since S has the homogeneity property, $S(\lambda V, \lambda E, \lambda N) = \lambda S(V, E, N)$, we have

$$S = V \left(\frac{\partial S}{\partial V}\right) + E \left(\frac{\partial S}{\partial E}\right) + N \left(\frac{\partial S}{\partial N}\right)$$

or

$$\left(\frac{\partial S}{\partial N}\right) = \frac{1}{N} \left[S - V \left(\frac{\partial S}{\partial V}\right) - E \left(\frac{\partial S}{\partial E}\right) \right] .$$

For homogeneous substances, $dS = (dE + p\,dV)/T$; that is,

$$\partial S/\partial V = p/T \quad \text{and} \quad \partial S/\partial E = 1/T .$$

Therefore,

$$\left(\frac{\partial S}{\partial N}\right) = \frac{1}{N} \left[S - \frac{pV}{T} - \frac{E}{T} \right] = -\frac{\Phi}{NT} .$$

Consequently, from Eq. [20.2],

$$\Phi_1/N_1 T_1 = \Phi_2/N_2 T_2 \quad \text{or} \quad \Phi_1/N_1 = \Phi_2/N_2 ;$$

that is,

$$p_1 = p_2 .$$

The above relations are necessary for stable equilibrium,

but they are not sufficient. Therefore, we still must investigate the second variation. We must have

$$\delta^2 S = \sum_i \delta^2 S_i = 2 \sum_i \left\{ \frac{1}{2} \left(\frac{\partial^2 S_i}{\partial E_i^2} \right) (dE_i)^2 \right.$$
$$\left. + \left(\frac{\partial^2 S_i}{\partial E_i \partial N_i} \right) dE_i \, dN_i + \frac{1}{2} \left(\frac{\partial^2 S_i}{\partial N_i^2} \right) (dN_i)^2 \right\} \leqslant 0 ,$$

with the subsidiary conditions

$(dE_1)^2 = (dE_2)^2 , \quad dE_1 \, dN_1 = dE_2 \, dN_2 , \text{ and } (dN_1)^2 = (dN_2)^2 .$

Since in the equilibrium state the quantities

$$N_i \frac{\partial^2 S_i}{\partial E_i^2} , \quad N_i \frac{\partial^2 S_i}{\partial E_i \partial N_i} , \quad \text{and} \quad N_i \frac{\partial^2 S_i}{\partial N_i^2}$$

are independent of i, we can introduce the following quantities: internal energy per mole, $e = E/N = E_i/N_i$; molar volume, $v = V/N = V_i/N_i$; and entropy per mole, $s(e, v) = S/N = S_i/N_i$. Then we obtain

$$\delta^2 S = 2N \left\{ \frac{1}{2} \left(\frac{\partial^2 s}{\partial e^2} \right) (de)^2 + \left(\frac{\partial^2 s}{\partial e \partial v} \right) de \, dv + \frac{1}{2} \left(\frac{\partial^2 s}{\partial v^2} \right) (dv)^2 \right\} \leqslant 0 .$$

In what follows we shall again write S, E, and V; regarding that, we must bear in mind that these quantities refer to one mole. With the new notation, the stability condition is

$$\delta^2 S = 2 \left\{ \frac{1}{2} \left(\frac{\partial^2 S}{\partial E^2} \right) (dE)^2 + \left(\frac{\partial^2 S}{\partial E \partial V} \right) dE \, dV + \frac{1}{2} \left(\frac{\partial^2 S}{\partial V^2} \right) (dV)^2 \right\} \leqslant 0.$$

We want to write this expression in terms of the free energy F:

$$S = - \left(\frac{\partial F}{\partial T} \right)_v , \qquad E = F - T \left(\frac{\partial F}{\partial T} \right)_v ,$$

$$dS = - \left(\frac{\partial^2 F}{\partial T^2} \right) dT - \left(\frac{\partial^2 F}{\partial T \partial V} \right) dV ,$$

$$dE = - T \left(\frac{\partial^2 F}{\partial T^2} \right) dT + \left[\left(\frac{\partial F}{\partial V} \right)_T - T \left(\frac{\partial^2 F}{\partial T \partial V} \right) \right] dV .$$

Therefore,

$$(\mathrm{d}V = 0), \quad \left(\frac{\partial S}{\partial E}\right)_V = \frac{1}{T}; \quad \text{and} \quad (\mathrm{d}E = 0), \quad \left(\frac{\partial S}{\partial V}\right)_E = -\frac{1}{T}\left(\frac{\partial F}{\partial V}\right)_T .$$

These relations follow from

$$\left(\frac{\partial S}{\partial V}\right)_E = -\left(\frac{\partial^2 F}{\partial T^2}\right)\left(\frac{\partial T}{\partial V}\right)_E - \frac{\partial^2 F}{\partial T \partial V}$$

and

$$\left(\frac{\partial^2 F}{\partial T^2}\right)\left(\frac{\partial T}{\partial V}\right)_E = \frac{1}{T}\left(\frac{\partial F}{\partial V}\right)_T - \left(\frac{\partial^2 F}{\partial T \partial V}\right).$$

Using the relations

$$\mathrm{d}\left(\frac{\partial S}{\partial E}\right) = \left(\frac{\partial^2 S}{\partial E^2}\right)\mathrm{d}E + \left(\frac{\partial^2 S}{\partial E \partial V}\right)\mathrm{d}V$$

and

$$\mathrm{d}\left(\frac{\partial S}{\partial V}\right) = \left(\frac{\partial^2 S}{\partial V \partial E}\right)\mathrm{d}E + \left(\frac{\partial^2 S}{\partial V^2}\right)\mathrm{d}V ,$$

we can write

$$
\begin{aligned}
\delta^2 S &= \mathrm{d}\left(\frac{\partial S}{\partial E}\right)\mathrm{d}E + \mathrm{d}\left(\frac{\partial S}{\partial V}\right)\mathrm{d}V \\
&= \mathrm{d}\left(\frac{1}{T}\right)\left\{-T\left(\frac{\partial^2 F}{\partial T^2}\right)\mathrm{d}T + \left[\left(\frac{\partial F}{\partial V}\right) - T\left(\frac{\partial^2 F}{\partial T \partial V}\right)\right]\mathrm{d}V\right\} \\
&\quad + \mathrm{d}\left[-\frac{1}{T}\left(\frac{\partial F}{\partial V}\right)\right]\mathrm{d}V \\
&= \left\{\frac{\mathrm{d}T}{T}\left(\frac{\partial^2 F}{\partial T^2}\right)\mathrm{d}T - \frac{\mathrm{d}T}{T^2}\left[\left(\frac{\partial F}{\partial V}\right) - T\left(\frac{\partial^2 F}{\partial T \partial V}\right)\right]\mathrm{d}V\right\} \\
&\quad + \left\{\left[\frac{\mathrm{d}T}{T^2}\left(\frac{\partial F}{\partial V}\right) - \frac{1}{T}\left(\frac{\partial^2 F}{\partial T \partial V}\right)\mathrm{d}T\right]\mathrm{d}V - \frac{1}{T}\left(\frac{\partial^2 F}{\partial V^2}\right)(\mathrm{d}V)^2\right\} \\
&= \frac{1}{T}\left\{\left(\frac{\partial^2 F}{\partial T^2}\right)(\mathrm{d}T)^2 - \left(\frac{\partial^2 F}{\partial V^2}\right)(\mathrm{d}V)^2\right\} \\
&= -\frac{1}{T}\left\{\frac{c_v}{T}(\mathrm{d}T)^2 - \left(\frac{\partial p}{\partial V}\right)_T(\mathrm{d}V)^2\right\} \leqslant 0 .
\end{aligned}
$$

From this we obtain the following equilibrium conditions:

for $V = $ constant, $c_V \geqslant 0$; and, for $T = $ constant, $(\partial p / \partial V)_T \leqslant 0$.

21. THERMOELECTRIC PROBLEMS

Because of irreversible processes, these effects are partially removed from the scope of thermodynamics. A statistical explanation is possible; however, it goes beyond the framework of phenomenological thermodynamics.

a. Thomson's effect

If an electric current J flows through a wire, between the ends of which there is a temperature difference ΔT, then an amount of heat Q is produced per second, given by

$$Q = \tau \Delta T J .$$

Depending on the material, τ can be positive or negative. This process is reversible.

b. Peltier's effect

At the boundary between two metals through which a current is passing, the heat $Q = \pi J$ is produced per second. The Peltier constant π depends on the temperature T.

c. Thermal emf

In the solder joint between two metals, an emf,

$$E_{12}(T) = E_2 - E_1 ,$$

is produced.

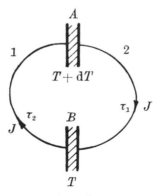

Figure 21.1

Let two different wires be soldered to one another at their ends A and B. Joint A is to be at temperature $T+dT$, and joint B at temperature T. In A the emf $E_{12}+dE_{12}$ is produced, and in B the emf E_{21} is produced. In A the Peltier heat $\pi+d\pi$ is absorbed per unit current, and in B, π is given up per unit current. The amount of heat produced per unit current in the two wires because of the Thomson effect is $(\tau_2-\tau_1)dT$. Since the "internal energy" of the system does not change, therefore, according to the first law, the amount of work done per unit current must equal the amount of heat absorbed per unit current. Since the amount of work done per unit current per second equals the potential difference between A and B, we have

$$dE = d\pi + (\tau_2-\tau_1)\,dT\,,$$

or

$$\frac{dE}{dT} = \frac{d\pi}{dT} + (\tau_2-\tau_1)\,.$$

Integrating from T_1 to T_2, we obtain

$$E(T_2) - E(T_1) = \pi(T_2) - \pi(T_1) + \int_{T_1}^{T_2}(\tau_2-\tau_1)\,dT\,.$$

If we view the entire process as reversible, which is not exactly correct, we can apply the second law and obtain

$$\frac{\pi(T_2)}{T_2} - \frac{\pi(T_1)}{T_1} + \int_{T_1}^{T_2}(\tau_2-\tau_1)\frac{dT}{T} = 0\;;$$

this follows from

$$\oint(\delta Q)/T = 0\,.$$

Since T_1 and T_2 are nearly equal temperatures, we may differentiate and obtain

$$\frac{d}{dT}\left(\frac{\pi}{T}\right) + \frac{\tau_2-\tau_1}{T} = 0\,.$$

Taken together with the result of the first law, we have

$$\frac{d}{dT}\left(\frac{\pi}{T}\right) + \frac{1}{T}\left[\frac{dE}{dT} - \frac{d\pi}{dT}\right] = 0$$

or

$$-\left(\frac{\pi}{T^2}\right) + \frac{1}{T}\left(\frac{dE}{dT}\right) = 0 \ .$$

This yields precisely the relation

$$\frac{\pi}{T} = \frac{dE}{dT},$$

which is due to Thomson. Also,

$$(\tau_2 - \tau_1) = -T\frac{d}{dT}\left(\frac{\pi}{T}\right) = -T\left(\frac{d^2 E}{dT^2}\right).$$

The derivation of this relation is, as a matter of fact, not correct; nevertheless, the result is correct, as one can determine on the basis of statistical considerations.[3] Following Boltzmann we can, with thermodynamics, rigorously derive [4]

$$\left[\frac{dE}{dT} - \frac{\pi}{T}\right]^2 \leqslant \frac{4}{T}\overline{\left(\frac{\lambda}{\sigma}\right)},$$

where

$$\overline{\left(\frac{\lambda}{\sigma}\right)} \equiv \left[\sqrt{\frac{\lambda_1}{\sigma_1}} + \sqrt{\frac{\lambda_\upsilon}{\sigma_2}}\right]^2,$$

λ = heat conductivity and σ = electrical conductivity.

[3] L. ONSAGER, *Phys. Rev.* **37**, 405 (1931) and *Phys. Rev.* **38**, 2265 (1931). Also, H. B. G. CASIMIR, *Rev. Mod. Phys.* **17**, 343 (1945).
[4] L. BOLTZMANN, *Wien. Ber.* **96**, 1258 (1887).

Chapter 4. Nernst's Heat Theorem

22. NERNST'S HEAT THEOREM

The Nernst heat theorem is concerned with the behavior of the thermodynamic function in the neighborhood of the absolute zero. It is well known that the second law determines the entropy only to within an additive constant. This constant is sensibly normalized by means of the Nernst heat theorem.

Let us first consider the free energy:

$$F = E + T\left(\frac{\partial F}{\partial T}\right), \qquad \Delta F = \Delta E + T\frac{\partial(\Delta F)}{\partial T}.$$

From this it follows that

$$\lim_{T \to 0}(\Delta F - \Delta E) = 0.$$

Nernst requires the stronger condition

$$\lim_{T \to 0}\frac{\Delta F - \Delta E}{T} = 0.$$

From this follow the equivalent relations

$$\lim_{T \to 0}\frac{\partial(\Delta F)}{\partial T} = 0, \qquad \lim_{T \to 0}\frac{\partial(\Delta \Phi)}{\partial T} = 0,$$

and further,

$$\lim_{T \to 0}\Delta S = 0.$$

This is the more restricted formulation of the Nernst theorem; it says that all entropy changes are zero at the absolute zero. A generalization of the Nernst theorem due to Planck states that in the relation $\lim_{T\to 0}\Delta S=0$, the Δ can be dropped; $S_0=\lim_{T\to 0}S$ should be finite and universal. A reasonable normalization of the entropy constant is such that

$$\lim_{T\to 0} S = 0 \; ;$$

that is, the entropy of all pure substances has the value zero at the absolute zero, $T=0$. From this requirement it follows that the entropy of all substances must be zero at the absolute zero. This is because the entropy change ΔS which occurs when a substance is made from pure substances at absolute zero is zero according to Nernst.

Because $\lim_{T\to 0}S=0$, it follows that

$$\lim_{T\to 0}\left(\frac{\partial S}{\partial V}\right)_T = 0 \quad \text{and} \quad \lim_{T\to 0}\left(\frac{\partial S}{\partial p}\right)_T = 0 \, .$$

From this it follows that

$$\lim_{T\to 0}\left(\frac{\partial^2 F}{\partial T\,\partial V}\right) = 0 \quad \text{and} \quad \lim_{T\to 0}\left(\frac{\partial p}{\partial T}\right)_V = 0 \, .$$

Likewise,

$$\lim_{T\to 0}\left(\frac{\partial^2 \Phi}{\partial T\,\partial p}\right) = 0 \quad \text{and} \quad \lim_{T\to 0}\left(\frac{\partial V}{\partial T}\right)_p = 0 \, .$$

From this it further follows that [1]

$$\lim_{T\to 0} (c_p - c_V) = 0 \; ;$$

that is, for $T=0$, we must have $c_p=c_v$. From

$$c_V = T\left(\frac{\partial S}{\partial T}\right)_V, \quad c_p = T\left(\frac{\partial S}{\partial T}\right)_p, \quad \text{and} \quad \lim_{T\to 0} S = 0$$

[1] See Eq. [12.7].

follows

$$S = \int\limits_0^T \frac{c_V}{T}\,\mathrm{d}T \quad \text{or} \quad S = \int\limits_0^T \frac{c_p}{T}\,\mathrm{d}T \ .$$

In its generalized formulation the Nernst heat theorem states that the above two integrals exist. If $\lim\limits_{T \to 0} c_V$ exists, then, for $T=0$, we must have $c_p = c_V = 0$. Since

$$S = -(\partial F/\partial T)_V \quad \text{and} \quad S = -(\partial \Phi/\partial T)_p,$$

it follows that

$$F = -\int\limits_0^T \mathrm{d}T' \int\limits_0^T \frac{c_V(T'', V)}{T''}\,\mathrm{d}T'' + E_0(V)$$

and

$$\Phi = -\int\limits_0^T \mathrm{d}T' \int\limits_0^{T'} \frac{c_p(T'', p)}{T''}\,\mathrm{d}T'' + E_0 + pV_0 \ .$$

23. UNATTAINABILITY OF THE ABSOLUTE ZERO

The requirement of the unattainability of the absolute zero is contained in Nernst's theorem; it is, however, weaker than Nernst's theorem. In order to prove the unattainability, it is sufficient to show that it is impossible to reach the absolute zero by adiabatic changes of state, because every process can be decomposed into adiabatic and isothermal processes. According to Nernst's theorem, there does not exist an adiabatic curve along which one could reach the absolute zero. Since at the absolute zero the adiabatic curve $S=0$ is the same as the isotherm $T=0$, therefore no adiabatic curve $S=\text{constant} \neq 0$ can intersect the adiabatic curve $S=0$, which corresponds to the isotherm $T=0$. Thus, it is not possible to reach the absolute zero by means of adiabatic changes of state.

For ideal gases the entropy per mole is

$$S = c_V \log T + R \log V + a = R \log \left(V T^{\frac{c_V}{R}} \right) + a \, .$$

At $T=0$, S does not go through zero for finite changes of state. It would appear that Nernst's theorem is satisfied. However, because of

$$\lim_{T \to 0} \left[S(T, V_2) - S(T, V_1) \right] = R \log \left(\frac{V_2}{V_1} \right) \neq 0 \, ,$$

this is not true. In spite of the fact that in this case the absolute zero is not attainable, Nernst's theorem is not satisfied. One must not doubt the Nernst theorem because of this. Rather, one must accept it as correct and, at the same time, assume that the above equation for ideal gases is no longer valid at very low temperatures (in the neighborhood of the absolute zero); that is, the ideal gases become degenerate. The quantum theory confirms this assumption.

Chapter 5. Kinetic Theory of Gases

In connection with the atomic structure of matter, it must be assumed that the smallest components of matter, that is, atoms and molecules, are not at rest but are in motion. Of course, this motion cannot be seen directly; however, its existence is apparent from Brownian motion. Because of this, first Krönig, and later Maxwell and Clausius, formed the following hypothesis: Heat energy is identical with the kinetic energy of the molecules or atoms. Therefore, one can say that heat is a disordered form of energy. Boltzmann introduced the concept of the "probability of a state" and connected it with entropy. The law that the entropy can only increase is identical with the statement that a system which changes its state can only go to a more probable state.

The various states of matter are differentiated as follows:

Gases: Except for collisions, the molecular motion is force free (neglecting external force fields).

Solids: The atoms oscillate about an equilibrium configuration.

Liquids: Because there are always a large number of molecules which interact, neither uniform motion nor an equilibrium configuration is a good approximation.

Today the theory of gases and the theory of the structure of the solid state are very advanced, in contrast with the theory of liquids, which is not yet well founded.

With ideal gases we may make the following assumptions:

1. The average separation of two free molecules is very large compared with the molecular dimensions.

2. The average potential energy of the molecules is very small compared with the kinetic energy and, therefore, we can neglect it.

We shall assume that a gas consists of hard spheres. It turns out, however, that the majority of the results are independent of the specific model.

24. CALCULATION OF THE PRESSURE

We want to calculate the pressure which the molecules of an ideal gas exert on a completely elastic wall. Let $f(v_1, v_2, v_3) \, dv_1 dv_2 dv_3$ be that fraction of the molecules for which the velocity (v_1, v_2, v_3) lies in the interval v_1 to $v_1 + dv_1$, v_2 to $v_2 + dv_2$, v_3 to $v_3 + dv_3$. Let f be so normalized that

$$\int f \, \mathrm{d}^3 v = 1 \, ,$$

where

$$\mathrm{d}^3 v = \mathrm{d}v_1 \, \mathrm{d}v_2 \, \mathrm{d}v_3 \, .$$

The pressure is given by the momentum delivered to the wall per unit time and per unit area. The molecules for which the velocity lies in the interval v_1 to $v_1 + dv_1$, v_2 to $v_2 + dv_2$, v_3 to $v_3 + dv_3$, and which hit 1 cm² of the wall in a second, are in a cylinder of volume v_1 (see Fig. 24.1). Therefore, the number of these molecules is $n v_1 f(v_1, v_2, v_3) \mathrm{d}^3 v$, where n is the number of molecules per cubic centimeter. Momentum $2mv_1$ is delivered to the wall by one molecule; therefore, momentum $2m v_1^2 n f(v_1, v_2, v_3) \, \mathrm{d}^3 v$ is delivered by all of the molecules in the cylinder. Thus, the pressure p

has the value

$$p = 2mn \int\limits_{v_1 > 0} v_1^2 f(v_1, v_2, v_3) \, \mathrm{d}^3 v \, .$$

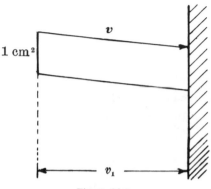

Figure 24.1

If we assume that f is an even function in v_1, that is, $f(-v_1, v_2, v_3) = f(v_1, v_2, v_3)$, then

$$p = mn \int v_1^2 f(v_1, v_2, v_3) \, \mathrm{d}^3 v \, .$$

We define $\overline{v_1^2}$, the average value of v_1, as

$$\overline{v_1^2} = \int v_1^2 f(v_1, v_2, v_3) \, \mathrm{d}^3 v \, .$$

With this definition, we have

$$p = mn\overline{v_1^2} \, .$$

This result is independent of the nature of the wall. If an arbitrary wall is given, then, by adding an imaginary surface, we make it into a closed surface.[1] We can define a momentum tensor

$$T_{ik} = nm\overline{v_i v_k} \, .$$

[1] See Fig. 24.2.

The ith component of the momentum transported per unit time across the imaginary surface F is then

$$\int_F \sum_k T_{ik} \cos (N, \widehat{x}_k) \, \mathrm{d}f,$$

where N is the outward normal to $\mathrm{d}f$.

An analogous expression, in which the pressure appears

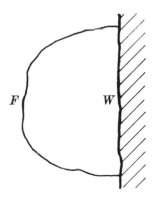

Figure 24.2

as a tensor p_{ik}, is valid for the wall W. Considering the balance of momentum within the volume bounded by F and W, we obtain

$$\int_F \sum_k T_{ik} \cos (N, \widehat{x}_k) \, \mathrm{d}f + \int_W \sum_k p_{ik} \cos (N, \widehat{x}_k) \, \mathrm{d}f$$

$$+ \frac{\mathrm{d}}{\mathrm{d}t} \int nm\overline{v}_i \, \mathrm{d}V = 0. \qquad [24.1]$$

If we substitute $A = a\varphi$ ($a = $ constant) into Gauss's theorem,

$$\int A \cdot \mathrm{d}f = \int \mathrm{div}\, A \, \mathrm{d}V,$$

then, using $\operatorname{div} \boldsymbol{A} = \boldsymbol{a} \cdot \operatorname{grad} \varphi$, we obtain

$$\int \boldsymbol{a}\varphi \cdot \mathrm{d}\boldsymbol{f} = \boldsymbol{a} \cdot \int \operatorname{grad} \varphi \, \mathrm{d}V ,$$

$$\sum_k a_k \int \varphi \cos{(\boldsymbol{N}, \hat{x}_k)} \, \mathrm{d}f = \sum_k a_k \int \frac{\partial \varphi}{\partial x_k} \, \mathrm{d}V ,$$

$$\int \varphi \cos{(\boldsymbol{N}, \hat{x}_k)} \, \mathrm{d}f = \int \frac{\partial \varphi}{\partial x_k} \, \mathrm{d}V .$$

Substituting T_{ik} for φ and letting F represent a closed surface, we obtain from Eq. [24.1]

$$\int \sum_k \frac{\partial T_{ik}}{\partial x_k} \, \mathrm{d}V + \frac{\mathrm{d}}{\mathrm{d}t} \int \varrho \bar{v}_i \, \mathrm{d}V = 0 ,$$

where

$$\varrho = nm .$$

Hence

$$\frac{\mathrm{d}}{\mathrm{d}t} (\varrho \bar{v}_i) = - \sum_k \frac{\partial T_{ik}}{\partial x_k} \qquad \text{(hydrodynamic equation)} .$$

For the stationary case, from this equation as well as from Eq. [24.1], we obtain [A-12]

$$p_{ik} = T_{ik} \qquad \text{(independent of position)} .$$

In the case of an isotropic velocity distribution, we have

$$p_{ik} = \delta_{ik} p \qquad \text{and} \qquad \overline{v_i v_k} = \delta_{ik} (\overline{v^2}/3);$$

therefore,

$$p = \tfrac{1}{3} nm \overline{v^2} .$$

The kinetic energy of a single molecule is $E_{\text{kin}} = \tfrac{1}{2} mv^2$, and the kinetic energy of all of the molecules is $U = \tfrac{1}{2} mn \overline{v^2}$; therefore,

$$p = 2U/3 .$$

If v is the molar volume and L is Avogadro's number, then $L = nv$, and we obtain

$$pv = \tfrac{2}{3} L \bar{E}_{\text{kin}} .$$

According to Boyle's law, $pv = RT$; from this it follows that

$$\bar{E}_{\text{kin}} = \frac{1}{2}m\bar{v^2} = \frac{3R}{2L}T = \frac{3}{2}kT\,,$$

where $k = R/L$ is Boltzmann's constant. The average kinetic energy is thus a function of temperature. Here we can make the following important new assumption: In thermodynamic equilibrium the average kinetic energies of two molecules are equal to one another. Under the assumption that the kinetic energy accounts for the entire heat energy, we obtain

$$c_V = \frac{3}{2}R \sim 3\,\frac{\text{cal}}{\text{mole}}$$

for the specific heat. The second assumption is not generally valid. It is valid only for monatomic gases (noble gases, metal vapors). For polyatomic gases, $c_V > 3R/2$. This means that the kinetic energy does not account for the total heat energy. Strictly speaking, these statements are outside the realm of the kinetic theory of gases. Since there are three degrees of freedom associated with the translational motion, and since the kinetic energy in this case (monatomic gases) is $\bar{E}_{\text{kin}} = 3kT/2$, it is assumed that the kinetic energy per degree of freedom is $kT/2$. In the case of polyatomic gases the rotational energy must be considered along with the translational energy.

Monatomic gases: $\quad f = 3\,, \quad c_V = 3R/2\,, \quad \varkappa = 1.66\,;$

diatomic gases: $\quad f = 5\,, \quad c_V = 5R/2\,, \quad \varkappa = 1.40\,;$

triatomic gases: $\quad f = 6\,, \quad c_V = 3R\,, \quad \varkappa = 1.33\,.$

The quantity $\varkappa$ is calculated from the following relations:

$$c_V = \frac{f}{2}R\,, \quad c_p - c_V = R\,, \quad \varkappa = \frac{c_p}{c_V} = 1 + \frac{R}{c_V} = \frac{f+2}{f}\,.$$

The difficulties are associated with the determination of the number of degrees of freedom. (For details see the lectures on statistical mechanics.)

The value of the average speed $\bar{v} = \sqrt{\overline{v^2}}$ of a gas molecule is of the order of magnitude of the speed of sound V:

$$V = \sqrt{\varkappa \frac{p}{\varrho}} = \sqrt{\varkappa \frac{RT}{M}} \,.$$

From

$$\frac{m}{2}\,\overline{v^2} = \frac{3}{2}\,kT$$

follows

$$\bar{v} = \sqrt{\overline{v^2}} = \sqrt{3\,\frac{RT}{M}} = \sqrt{\frac{3}{\varkappa}}\,V \,.$$

25. VELOCITY DISTRIBUTION

The exact derivation of the velocity distribution proceeds from a consideration of the collisions of the individual molecules or atoms. In the case of an isotropic velocity distribution, we have

$$f(v_1, v_2, v_3)\,\mathrm{d}^3 v = f(v^2)\,\mathrm{d}^3 v = f(v^2)v^2\,\mathrm{d}v\,\mathrm{d}\Omega,$$

where $v^2 = v_1^2 + v_2^2 + v_3^2$, and where $\mathrm{d}\Omega$ is the differential solid angle. (In polar coordinates $\mathrm{d}^3 v = v^2\mathrm{d}v\,\mathrm{d}\Omega$.) Maxwell postulated that the distribution of one component of the velocity is independent of the other components. Therefore, he set $f(v^2) = g(v_1)g(v_2)g(v_3)$. Introducing the functions $\psi(v_i)$ and $\varphi(v^2)$ such that

$$\log g(v_i) = \psi(v_i) \qquad (i = 1, 2, 3)$$

and

$$\log f(v^2) = \varphi(v^2)\,,$$

we obtain the functional equation

$$\varphi(v^2) = \varphi(v_1^2 + v_2^2 + v_3^2) = \psi(v_1) + \psi(v_2) + \psi(v_3)\,.$$

A solution exists only if $\varphi(v^2)$ and $\psi(v_i)$ are linear functions of v_i^2; that is,

$$\psi(v_i^2) = -\alpha v_i^2 + \beta \quad \text{or} \quad g(v_i) = \text{constant} \times e^{-\alpha v_i^2}.$$

From this it follows that

$$f(v^2) = \text{constant} \times e^{-\alpha(v_1^2 + v_2^2 + v_3^2)} = \text{constant} \times e^{-\alpha v^2}.$$

The constant is determined from the normalization $\int\limits_v f \, dv = 1$. From

$$\int\limits_{-\infty}^{+\infty} e^{-\alpha v_i^2} \, dv_i = \frac{1}{\sqrt{\alpha}} \int\limits_{-\infty}^{+\infty} e^{-x^2} \, dx = \sqrt{\frac{\pi}{\alpha}},$$

we obtain

$$f(v^2) = \left(\frac{\alpha}{\pi}\right)^{\frac{3}{2}} e^{-\alpha v^2}.$$

Now, α can be determined from the mean square velocity $\overline{v_1^2}$, since $\overline{v_1^2} = kT/m$:

$$\overline{v_1^2} = \frac{\int\limits_{-\infty}^{+\infty} v_1^2 e^{-\alpha v_1^2} \, dv_1}{\int\limits_{-\infty}^{+\infty} e^{-\alpha v_1^2} \, dv_1} = \frac{\dfrac{1}{2\alpha} \int\limits_{-\infty}^{+\infty} v_1 (2\alpha v_1 e^{-\alpha v_1^2}) \, dv_1}{\int\limits_{-\infty}^{+\infty} e^{-\alpha v_1^2} \, dv_1} = \frac{1}{2\alpha};$$

hence

$$\alpha = \frac{m}{2kT}.$$

Finally, we obtain

$$f(v^2) \, d^3v = \left(\frac{m}{2\pi kT}\right)^{\frac{3}{2}} e^{-\left(\frac{m}{2kT}\right)v^2} d^3v$$

for the isotropic velocity distribution function. For the probability that a molecule has speed v, we obtain

$$w(v) \, dv = \int\limits_{\Omega} f(v^2) \, d^3v = \left(\frac{\alpha}{\pi}\right)^{\frac{3}{2}} e^{-\alpha v^2} v^2 \, dv \int\limits_{\Omega} d\Omega$$

$$= 4\pi \left(\frac{\alpha}{\pi}\right)^{\frac{3}{2}} v^2 e^{-\alpha v^2} dv = 4\pi \left(\frac{m}{2\pi kT}\right)^{\frac{3}{2}} v^2 e^{-\frac{m}{2kT}v^2} dv.$$

This yields the velocity distribution shown in Fig. 25.1. The most probable speed is obtained from the relation $dw(v)/dv = 0$:

$$1 - \alpha v_M^2 = 0 \qquad \text{or} \qquad v_M = \frac{1}{\sqrt{\alpha}}.$$

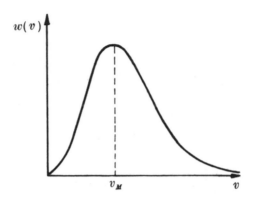

Figure 25.1. Maxwell velocity distribution.

Since the mean square speed is $\overline{v^2} = \sum_i \overline{v_i^2} = 3/(2\alpha)$, we obtain $v_M = (2\overline{v^2}/3)^{\frac{1}{2}}$. The average speed $\bar{v}$ is defined by

$$\bar{v} = \int_0^\infty v\, w(v)\, dv = 4\pi \left(\frac{\alpha}{\pi}\right)^{\frac{3}{2}} \int_0^\infty v^3\, e^{-\alpha v^2}\, dv\,.$$

After two partial integrations we obtain $\bar{v} = 2(\alpha\pi)^{-\frac{1}{2}}$. Therefore,

$$\bar{v} = \frac{2}{\sqrt{\pi}}\, v_M = \sqrt{\frac{8}{3\pi}\, \overline{v^2}}\,.$$

We now want to calculate the molecular current J which passes across a unit surface area in the x direction. With an arbitrary isotropic velocity distribution, the number of molecules with speed v which cross the unit surface area per second is

$$nv_x f(v)\, \mathrm{d}^3 v = nv_x f(v) v^2\, \mathrm{d}v\, \mathrm{d}\Omega\,,$$

where $\mathrm{d}\Omega = 2\pi \sin\vartheta\, \mathrm{d}\vartheta$. Therefore,

$$J = n\int\limits_0^{\pi/2} \cos\vartheta\, 2\pi \sin\vartheta\, \mathrm{d}\vartheta \int\limits_0^\infty v^3 f(v)\, \mathrm{d}v = \frac{n}{4}\,\bar{v}\,.$$

The Maxwell distribution can be experimentally verified with the experiment of Stern. Stern produced a molecular beam in a highly evacuated tube. This beam hits a screen

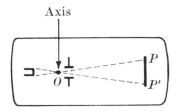

Figure 25.2. Aperture dimensions small compared with the free path.

at point P. If the tube is now allowed to rotate about an axis perpendicular to the molecular beam, then the beam is deflected due to the Coriolis force and strikes the screen at point P' instead of at point P. The velocity distribution of the molecular beam can be determined from the deflection $\overline{PP'}$.

26. MEAN FREE PATH AND COLLISIONS

We imagine that the gas consists of hard spheres. Then two molecules will collide when the center of one comes within the interaction sphere of the other. The interaction sphere of a molecule is the sphere whose radius is $\sigma = 2r$, twice that of the molecule. If we take force centers instead of hard spheres, a quantity σ can be defined in a like manner; the quantity is velocity-dependent in that case. Let w be the relative velocity of two colliding molecules with velocities v and v'; and let V be the center of mass velocity of the

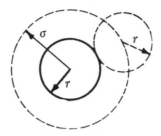

Figure 26.1

system consisting of these two molecules. Then, under the assumption that the two molecules have equal masses,

$$w_k = v_k - v_k' \qquad (k = 1, 2, 3)$$

and

$$V_k = \tfrac{1}{2}(v_k + v_k').$$

The number of collisions Z per unit time (one second) is

$$Z = \pi\sigma^2 \overline{w} n,$$

because all molecules which are in the cylinder drawn in Fig. 26.2 collide with a given molecule; $\overline{w}$ is the average

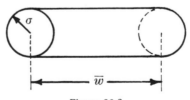

Figure 26.2

value of the relative speed w. The probability that two molecules with velocities v and v', respectively, hit one another is

$$f(\boldsymbol{v}, \boldsymbol{v}')\, \mathrm{d}^3 v\, \mathrm{d}^3 v' = f(v^2)\, \mathrm{d}^3 v\, f(v'^2)\, \mathrm{d}^3 v' = \left(\frac{\alpha}{\pi}\right)^3 e^{-\alpha(v^2 + v'^2)}\, \mathrm{d}^3 v\, \mathrm{d}^3 v' \,.$$

If we introduce the velocities $\boldsymbol{w}$ and $\boldsymbol{V}$ instead of $\boldsymbol{v}$ and $\boldsymbol{v}'$, then

$$v_k = V_k + \tfrac{1}{2} w_k \,,$$
$$v^2 + v'^2 = 2V^2 + \tfrac{1}{2} w^2 \,,$$
$$v_k' = V_k - \tfrac{1}{2} w_k \,.$$

Because

$$\frac{\partial(V_k, w_k)}{\partial(v_k, v_k')} = 1 \,,$$

we have $\mathrm{d}^3 v\, \mathrm{d}^3 v' = \mathrm{d}^3 V\, \mathrm{d}^3 w$. Therefore,

$$f(\boldsymbol{V}, \boldsymbol{w})\, \mathrm{d}^3 V\, \mathrm{d}^3 w = \left(\frac{\alpha}{\pi}\right)^3 e^{-\alpha(2V^2 + \frac{1}{2} w^2)}\, \mathrm{d}^3 V\, \mathrm{d}^3 w$$

and

$$f(w^2)\, \mathrm{d}^3 w = \text{constant} \times e^{-\frac{\alpha}{2} w^2}\, \mathrm{d}^3 w \,.$$

We then obtain

$$\overline{w} = \frac{\displaystyle\int_0^\infty w^3\, e^{-\frac{\alpha}{2} w^2}\, \mathrm{d} w}{\displaystyle\int_0^\infty w^2\, e^{-\frac{\alpha}{2} w^2}\, \mathrm{d} w} \,.$$

Since

$$\overline{v} = \frac{\displaystyle\int_0^\infty v^3\, e^{-\alpha v^2}\, \mathrm{d} v}{\displaystyle\int_0^\infty v^2\, e^{-\alpha v^2}\, \mathrm{d} v} \,,$$

we obtain

$$\overline{w} = \sqrt{2}\, \overline{v}$$

by making the trivial transformation $w = \sqrt{2}\, v$. If the average speed is $\overline{v}$, the number of collisions is

$$Z = n\pi\sqrt{2}\sigma^2\overline{v} \,.$$

In case the masses of the molecules are different, we introduce the reduced mass μ:

$$\mu = \frac{mm'}{m + m'} \qquad \text{or} \qquad \frac{1}{\mu} = \frac{1}{m} + \frac{1}{m'} .$$

The center of mass velocity V is then

$$V_k = \frac{mv_k + m'v_k'}{m + m'} ,$$

and

$$w_k = v_k - v_k' .$$

Then,

$$\frac{m}{2} v^2 + \frac{m'}{2} v'^2 = \frac{1}{2} (m + m') V^2 + \frac{\mu}{2} w^2$$

and, also,

$$\frac{\partial(V_k, w_k)}{\partial(v_k, v_k')} = 1 .$$

Therefore,

$$f(V, w) \, d^3V \, d^3w = \left(\frac{\alpha}{\pi}\right)^3 e^{-\frac{1}{kT}\left(\frac{m}{2} v^2 + \frac{m'}{2} v'^2\right)} d^3v \, d^3v'$$

$$= \left(\frac{\alpha}{\pi}\right)^3 e^{-\frac{1}{kT}\left\{(m + m')\frac{V^2}{2} + \frac{\mu}{2} w^2\right\}} d^3V \, d^3w ,$$

and

$$f(w^2) \, d^3w = \text{constant} \times e^{-\frac{1}{kT}\frac{\mu}{2} w^2} d^3w = \text{constant} \times e^{-\frac{\mu}{m}\alpha w^2} d^3w .$$

We thus find

$$\overline{w} = \sqrt{\frac{m}{\mu}} \, \overline{v} = \sqrt{\frac{m'}{\mu}} \, \overline{v'} .$$

If we define the free path as that distance which a molecule traverses without colliding with another, then for the mean free path we obtain

$$l = \frac{\text{distance}}{\text{number of collisions per distance}} = \frac{\overline{v}t}{Zt} = \frac{\overline{v}}{Z} = \frac{1}{n\sigma^2\pi\sqrt{2}} .$$

Since σ depends on the temperature, $\eta \sim \sqrt{T}$ cannot be taken literally. For the case of hard spheres, $a = 0.998$; that is, a is nearly 1. However, these formulas are not restricted to the case of hard spheres.

b. Heat conduction

There is a gas between two plates which have temperatures T_1 and T_0. The resulting heat current w is propor-

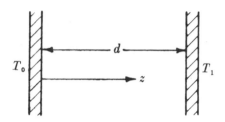

Figure 27.3

tional to the temperature gradient

$$\frac{\mathrm{d}T}{\mathrm{d}z} = \frac{T_1 - T_0}{d}.$$

According to the phenomenological theory,

$$w = -\varkappa \frac{\mathrm{d}T}{\mathrm{d}z},$$

where $\varkappa$ is the heat conductivity. Otherwise, the treatment is the same as before. We obtain, since $c = 0$, $v = u$,

$$w = -\frac{1}{2} n\bar{v}a'l \frac{\mathrm{d}E}{\mathrm{d}z}$$

($E =$ energy per molecule). If we introduce the quantity c_V as the specific heat per unit mass, then

$$\frac{\mathrm{d}E}{\mathrm{d}z} = mc_V \frac{\mathrm{d}T}{\mathrm{d}z};$$

from this it follows that

$$\varkappa = \tfrac{1}{2}\, nm\bar{v}a'lc_V\,.$$

Comparing the expressions for $\varkappa$ and η, we obtain

$$\varkappa = \frac{a'}{a}\, c_V\eta\,.$$

In the case of hard spheres, $a'/a = 2.5$.

c. Diffusion

The particle current per unit area is given by

$$i = -D\frac{\partial n}{\partial z}\,,$$

where D is the diffusion constant. In a way analogous to the previous considerations, we obtain

$$D = \tfrac{1}{2}a''\bar{u}l\,.$$

For gas mixtures,

$$D = a''\frac{1}{n}\,(n_1\bar{u}_1l_1 + n_2\bar{u}_2l_2)\,,$$

where $n = n_1 + n_2$.

All of these transport phenomena are independent of the density ϱ. However, the phenomenological theory is invalid for very low densities. In our previous considerations we tacitly assumed $d \gg l$. However, in the neighborhood of the plate, there is always a layer, whose thickness is of the order of l, in which $c_x(z) = \bar{c}z/d$ is no longer valid. Likewise, the calculation of the average velocity is incorrect for that layer. The distribution of velocities is then as shown in Fig. 27.4. A better approximation of the velocity gradient is then

$$\frac{\partial c_x}{\partial z} = \frac{\bar{c}}{d + 2\gamma}\,,$$

where $\gamma = \alpha l$ and α is a numerical coefficient. Thus γ is proportional to the free path. In the case of internal fric-

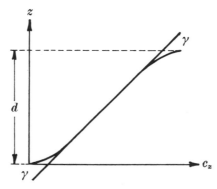

Figure 27.4

tion the shearing stresses are then

$$p_{zz} = -\eta \, \frac{\bar{c}}{d + 2\alpha l}.$$

For air at one atmosphere and 15 °C, $\alpha l \sim 10^{-5}$ cm. Analogously, in the case of heat conduction we must write

$$\frac{\mathrm{d}T}{\mathrm{d}z} = \frac{T_1 - T_0}{d + 2\alpha l}.$$

If $l \gg d$, we arrive at very simple results. Then there are practically no collisions of the particles with one another to take into account; there are only the collisions with the walls. Such a gas behaves like radiation. Heat exchange comes about only through collisions with the wall. There is no longer a temperature gradient. Then,

$$w = -\frac{\varkappa}{l}(T_1 - T_0) \sim -a'' c_V \varrho \bar{u}(T_1 - T_0).$$

The heat current is proportional to the density, but is independent of the separation of the plates. Knudsen carried out such experiments, which require a very high vacuum and very fine capillary tubes.

28. TRANSPORT PHENOMENA $(l \gg d)$

a. Diffusion through holes and pores

Let a gas-filled volume be divided into two parts by a wall. In the wall let there be an opening whose linear

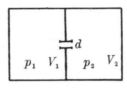

Figure 28.1

dimension is of the order of d. When $d \gg l$ and $p_1 = p_2$, there is no mass current through the opening. However, if $d \ll l$, then there is no mass current under the condition that

$$(n\bar{v})_1 = (n\bar{v})_2 .$$

Because $\bar{v} \sim \sqrt{T}$ and $n \sim \varrho$, this condition can be rewritten as

$$\varrho_1 \sqrt{T_1} = \varrho_2 \sqrt{T_2} ;$$

or, because $p \sim \varrho T$, it can also be expressed as

$$\frac{p_1}{\sqrt{T_1}} = \frac{p_2}{\sqrt{T_2}} \qquad \text{(condition for no mass current).}$$

b. Heat conduction at low pressures

Let us again consider the case of two plates at different temperatures T and T', now for the case $d \ll l$. Here, we need only consider the collisions with the plates and we can neglect the collisions between the molecules. We still must make an assumption concerning the energy exchange between the plates and the impinging molecules. We make the assumption that the energy which a molecule has following a collision with a plate is that which is determined

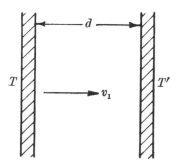

Figure 28.2

by the temperature of the plate. Therefore, the energy with which it hits the plate is not important. This ideal case is called "complete accommodation." Of the molecules which leave the plate (at temperature T) per unit time and per unit area, the fraction having velocity component v_1 is

$$v_1 f(v_1)\, dv_1 = \text{constant} \times v_1 e^{-\alpha v_1^2}\, dv_1\,.$$

The value of the constant, determined by the normalization $\int v_1 f(v_1)\,dv_1 = 1$, is 2α. The probability $w(v_1)\,dv_1$ that a molecule leaves the plate at temperature T with velocity component v_1 is

$$w(v_1)\, dv_1 = 2\alpha v_1 e^{-\alpha v_1^2}\, dv_1\,.$$

Analogously, the probability $w'(v_1)\,dv_1$ that a molecule leaves the plate at temperature T' with velocity component v_1 is

$$w'(v_1)\, dv_1 = 2\alpha' v_1 e^{-\alpha' v_1^2}\, dv_1\,.$$

A molecule with velocity component v_1 traverses a distance d in time d/v_1. Calculating the average value of this time, we obtain

$$\frac{1}{\nu} = \frac{1}{2} \left\{ \int\limits_0^\infty \frac{d}{v_1} w(v_1)\,dv_1 + \int\limits_0^\infty \frac{d}{v_1} w'(v_1)\,dv_1 \right\},$$

where $1/\nu$ is the average time which a molecule needs in order to traverse a distance d. Thus ν is the number of distances d that a molecule traverses per second. We can now write

$$\frac{1}{\nu} = d \left\{ \int\limits_0^\infty \alpha e^{-\alpha v_1^2}\,dv_1 + \int\limits_0^\infty \alpha' e^{-\alpha' v_1^2}\,dv_1 \right\} = \frac{d}{2}\sqrt{\pi}(\sqrt{\bar\alpha} + \sqrt{\bar\alpha'})$$

or

$$\nu = \frac{2}{d\sqrt{\pi}(\sqrt{\bar\alpha} + \sqrt{\bar\alpha'})}.$$

We now want to calculate the average heat transport per molecule emitted in the direction from (T) toward (T'). The distribution functions for the individual velocity components are

$$2\alpha v_1 e^{-\alpha v_1^2}\,dv_1, \quad \left(\frac{\alpha}{\pi}\right)^{\frac{1}{2}} e^{-\alpha v_2^2}\,dv_2, \quad \text{and} \quad \left(\frac{\alpha}{\pi}\right)^{\frac{1}{2}} e^{-\alpha v_3^2}\,dv_3.$$

Since a molecule carries an amount of energy $\frac{1}{2}mv^2 + \bar{E}_i$, where $\bar{E}_i$ is the average internal energy (rotational energy), we obtain

$$\int\limits_{v_1=0}^{+\infty} \int\limits_{-\infty}^{+\infty} \int\limits_{-\infty}^{+\infty} \left[\frac{m}{2}(v_1^2 + v_2^2 + v_3^2) + \bar{E}_i \right]$$

$$\times 2\alpha v_1 e^{-\alpha v_1^2}\,dv_1 \left(\frac{\alpha}{\pi}\right) e^{-\alpha v_2^2} e^{-\alpha v_3^2}\,dv_2\,dv_3$$

for the average heat transport per emitted molecule. In order to evaluate this integral, the following integrals are necessary:

$$\int\limits_{0}^{\infty} 2\alpha v_1^3 e^{-\alpha v_1^2}\, dv_1 = \int\limits_{0}^{\infty} 2v_1 e^{-\alpha v_1^2}\, dv_1 = \frac{1}{\alpha},$$

$$\int\limits_{-\infty}^{+\infty} v_i^2 e^{-\alpha v_i^2}\, dv_i = \left(\frac{\pi}{\alpha}\right)^{\frac{1}{2}} \frac{1}{2\alpha}.$$

In this way we obtain the following values for the heat transport per emitted molecule:

$\dfrac{m}{\alpha} + \bar{E}_i$ in the direction from (T) toward (T'),

$\dfrac{m}{\alpha'} + \bar{E}_i'$ in the direction from (T') toward (T).

A molecule traverses a distance d on the average v times per second and, indeed, $v/2$ times from (T) toward (T') and $v/2$ times from (T') toward (T). Therefore, a molecule transports an amount of heat

$$\frac{v}{2}\left(\frac{m}{\alpha} - \frac{m}{\alpha'} + \bar{E}_i - \bar{E}_i'\right)$$

per second from (T) toward (T'). Since there are $d \times n$ molecules per unit area between the two plates, the amount of heat transported per unit area per second is

$$\Delta w = dn \frac{v}{2}\left(\frac{m}{\alpha} - \frac{m}{\alpha'} + \bar{E}_i - \bar{E}_i'\right)$$

$$= \frac{n}{\sqrt{\pi}} \frac{1}{\sqrt{\bar{\alpha}} + \sqrt{\alpha'}} \left(\frac{m}{\alpha} - \frac{m}{\alpha'} + \bar{E}_i - \bar{E}_i'\right).$$

As was noted earlier, the separation of the plates drops out.

Since $\bar{E}_{\text{kin}} = m\overline{v^2}/2$ and $\overline{v_k^2} = 1/2\alpha \to \overline{v^2} = 3/2\alpha$, it follows that

$$\frac{m}{\alpha} = \frac{4}{3} \bar{E}_{\text{kin}}.$$

Therefore, we can write

$$\Delta w = \frac{n}{\sqrt{\pi}} \frac{1}{\sqrt{\alpha} + \sqrt{\alpha'}} \left\{ \left[\frac{4}{3} \bar{E}_{\text{kin}} + \bar{E}_i \right] - \left[\frac{4}{3} \bar{E}'_{\text{kin}} + \bar{E}'_i \right] \right\}$$

or, instead,

$$\Delta w = \frac{n}{\sqrt{\pi}} \frac{1}{\sqrt{\alpha} + \sqrt{\alpha'}} \left[\frac{4}{3} \frac{d\bar{E}_{\text{kin}}}{dT} + \frac{d\bar{E}_i}{dT} \right] (T - T') .$$

We define

$$c_V = \frac{1}{m} \left[\frac{d\bar{E}_{\text{kin}}}{dT} + \frac{d\bar{E}_i}{dT} \right]$$

to be the specific heat per unit mass. According to thermo-dynamics, we have

$$c_p - c_V = R$$

per mole; therefore, per unit mass,

$$c_p - c_V = \frac{R}{M} = \frac{k}{m} = \frac{2}{3m} \frac{d\bar{E}_{\text{kin}}}{dT} .$$

Because of this

$$c_p + c_V = \frac{2}{m} \left[\frac{4}{3} \frac{d\bar{E}_{\text{kin}}}{dT} + \frac{d\bar{E}_i}{dT} \right] ,$$

and we have

$$\Delta w = \frac{nm}{2\sqrt{\pi}} \frac{1}{\sqrt{\alpha} + \sqrt{\alpha'}} (c_p + c_V)(T - T') .$$

As a result of

$$nm = \varrho \quad \text{and} \quad \alpha = \frac{m}{2kT} = \frac{M}{2RT} = \frac{M}{2pv} = \frac{\varrho}{2p} ,$$

we have [3]

$$\Delta w = \frac{\varrho}{2\sqrt{\pi}} \frac{1}{\sqrt{\alpha} + \sqrt{\alpha'}} (c_p + c_V)(T - T')$$

$$\sim \frac{\varrho}{2\sqrt{\pi}} \frac{1}{2\sqrt{\alpha}} (c_p + c_V)(T - T') ,$$

$$\Delta w = \sqrt{\frac{p\varrho}{8\pi}} (c_p + c_V)(T - T') .$$

---

[3] For the general case, see H. A. LORENTZ, *Lectures on Theoretical Physics*, Vol. I: Kinetical Problems (Macmillan, London, 1927).

c. Flow through a tube at low pressures

Again let $l \gg a$, in order that we can regard the particle

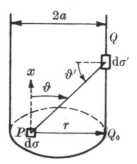

Figure 28.3

motion as radiation. The surface element $d\sigma$ radiates

$$A_P \cos\vartheta \, d\sigma \, d\Omega$$

toward Q [A-13], where $d\Omega = \sin\vartheta \, d\vartheta \, d\varphi$. Thus, $d\sigma$ radiates upward:

$$A_P \int\limits_0^{2\pi} d\varphi \int\limits_0^{\pi/2} \sin\vartheta \cos\vartheta \, d\vartheta \, d\sigma = A_P \pi \, d\sigma \ .$$

This is $A_P \pi$ per unit area of $d\sigma$, which must equal the molecular current per unit area in the x direction. That is,

$$A_P \pi = \frac{n}{4} \bar{v} \ .$$

Because

$$\bar{v} = \sqrt{\frac{8}{3\pi} \overline{v^2}} \quad \text{and} \quad \overline{v^2} = \frac{3}{2\alpha} = \frac{3p}{\varrho} = \frac{3p}{nm} \ ,$$

it follows that

$$A_P = \frac{p}{m\pi} \sqrt{\frac{3}{2\pi \overline{v^2}}} \ .$$

Here, as in photometry [A-13],

$$A_Q \cos\vartheta \, d\sigma \, d\Omega = A_Q \cos\vartheta' \, d\sigma' \, d\Omega'$$

applies to the mutual irradiation of two surface elements. However,

$$A_Q = A_P - \overline{QQ_0}\frac{dA_P}{dx} \; ;$$

and, because $\overline{QQ_0} = \overline{PQ}\cos\vartheta = r\cot\vartheta$,

$$A_Q = A_P - r\cot\vartheta \frac{dA_P}{dx} \, .$$

Therefore, the radiation from Q toward P is [A-13]

$$A_Q \cos\vartheta \, d\sigma \, d\Omega = A_P \cos\vartheta \, d\sigma \, d\Omega - r \frac{\cos^2\vartheta}{\sin\vartheta} \, d\sigma \, d\Omega \frac{dA_P}{dx}$$

$$= \left\{ A_P \cos\vartheta \sin\vartheta \, d\vartheta \, d\varphi - r\cos^2\vartheta \, d\vartheta \, d\varphi \frac{dA_P}{dx} \right\} d\sigma \, .$$

Integration over Ω yields [A-13]

$$-\frac{dA_P}{dx} \frac{\pi}{2} \, d\sigma \int_0^{2\pi} r \, d\varphi \; ;$$

integration over the cross section yields

$$J = -\frac{dA_P}{dx} \frac{\pi}{2} \int_\sigma d\sigma \int_0^{2\pi} r \, d\varphi$$

for the current. If the cross section is spherically symmetric, we obtain

$$\int_\sigma d\sigma \int_0^{2\pi} r \, d\varphi = \frac{16}{3} \, \pi a^3 \, ,$$

and therefore

$$J = -\frac{dA_P}{dx} \frac{8\pi^2}{3} \, a^3 = -\frac{dp}{dx} \frac{8a^3}{3m} \sqrt{\frac{3\pi}{2\overline{v^2}}} \, .$$

For J_m, the mass current, we obtain

$$J_m = -\frac{\mathrm{d}p}{\mathrm{d}x} \frac{8a^3}{3} \sqrt{\frac{3\pi}{2\overline{v^2}}} \,.$$

Comparing this formula with Poiseuille's formula,

$$J_p = -\frac{\mathrm{d}p}{\mathrm{d}x} \frac{\pi}{8} \frac{a^4\varrho}{\eta} \,,$$

we obtain

$$\frac{J_m}{J_p} = \text{constant} \times \frac{l}{a} \,.$$

29. VIRIAL CONCEPT

Let force $\boldsymbol{K}_i$ act on a particle of mass m_i. For an arbitrary number of particles the virial is defined as

$$w = \sum_i \boldsymbol{x}_i \cdot \boldsymbol{K}_i \,.$$

If $\sum_i \boldsymbol{K}_i = 0$, then it does not matter which point is chosen as the origin of the coordinate system. Since $\boldsymbol{K}_i = m_i \ddot{\boldsymbol{x}}_i$ and

$$m_i \boldsymbol{x}_i \cdot \ddot{\boldsymbol{x}}_i = m_i \frac{\mathrm{d}}{\mathrm{d}t} (\boldsymbol{x}_i \cdot \dot{\boldsymbol{x}}_i) - m_i \dot{\boldsymbol{x}}_i^2 \,,$$

therefore

$$w = \sum_i m_i \frac{\mathrm{d}}{\mathrm{d}t} (\boldsymbol{x}_i \cdot \dot{\boldsymbol{x}}_i) - \sum_i m_i \dot{\boldsymbol{x}}_i^2 \,.$$

Because the time average as well as the statistical average of an exact differential vanishes, we have

$$\overline{\frac{\mathrm{d}}{\mathrm{d}t} (\boldsymbol{x}_i \cdot \dot{\boldsymbol{x}}_i)} = 0 \,,$$

as long as $\boldsymbol{x}$ and $\dot{\boldsymbol{x}}$ are always finite. Therefore,

$$w = -\sum_i m_i \dot{\boldsymbol{x}}_i^2$$

or

$$w + \sum_i m_i \boldsymbol{v}_i^2 = 0 \qquad \text{(virial theorem)} \,.$$

30. APPLICATIONS

a. *Ideal gases*

In the case of an ideal gas in a volume V, $\boldsymbol{x}$ and $\dot{\boldsymbol{x}}$ are always finite. Per mole,

$$\sum_i m_i \boldsymbol{v}_i^2 = 2\bar{E}_{\text{kin}} = 3RT \ .$$

This expression is also correct if forces are assumed to act between the molecules. The virial of the pressure forces

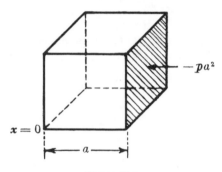

Figure 30.1

in the case of a cube is

$$w = -3pv \ .$$

This relation is valid for volumes of all shapes. For a cube the force on one side is

$$K = -pa^2 \ .$$

However, since for three of the sides $\boldsymbol{x} \cdot \boldsymbol{K} = aK$, and for the other three sides $\boldsymbol{x} \cdot \boldsymbol{K} = 0$, therefore

$$w = -3pa^3 = -3pv \ .$$

For a volume of arbitrary shape the virial can be calculated by means of Gauss's law:

$$w = -p \oint_{\text{surface}} \boldsymbol{x} \cdot \boldsymbol{n} \, \mathrm{d}f = -p \int_v (\mathrm{div}\, \boldsymbol{x}) \, \mathrm{d}v = -3p \int_v \mathrm{d}v = -3pv \, .$$

If no other forces need be considered, then per mole,

$$-3pv + 3RT = 0 \quad \text{or} \quad pv = RT \; ;$$

that is, we obtain exactly the ideal gas equation.

b. Real gases (Correction to the ideal gas equation)

Let there be forces between the molecules. Let the potential of the central forces be $U(r)$, and let it be so normalized that $U(\infty) = 0$. The forces $\boldsymbol{K_r}$ between two molecules are

$$\boldsymbol{K_r} = -\, \mathrm{grad}\, U(r) = -\frac{\mathrm{d}U}{\mathrm{d}r}\frac{\boldsymbol{x}}{r},$$

where

$$r = |\boldsymbol{x}| \quad \text{and} \quad \boldsymbol{x} = \boldsymbol{x_1} - \boldsymbol{x_2}$$

for $r > \sigma$, where σ is again the interaction sphere. For $r < \sigma$, we can set $U(r)|_{r<\sigma} = \infty$. The virial for the pair of molecules is then

$$\boldsymbol{x} \cdot \boldsymbol{K_r} = -\frac{\mathrm{d}U}{\mathrm{d}r}\frac{\boldsymbol{x}}{r} \cdot \boldsymbol{x} = -\frac{\mathrm{d}U}{\mathrm{d}r} r \, .$$

In order to obtain the virial for the interaction of one molecule with all others, we must integrate from σ to ∞. The volume element is $4\pi r^2 \, \mathrm{d}r$ and there are L/v molecules per cubic centimeter. (L is Avogadro's number, and v is the molar volume.) The result is

$$-\frac{L}{v} \int_\sigma^\infty \frac{\mathrm{d}U}{\mathrm{d}r} \, r 4\pi r^2 \, \mathrm{d}r \, .$$

In obtaining this result it was assumed that the particles interact only pairwise; that is, there are to be no many-body forces. Since there are L molecules per mole, we obtain

$$w = -\frac{L^2}{2v} \int_\sigma^\infty \frac{\mathrm{d}U}{\mathrm{d}r} 4\pi r^3 \mathrm{d}r$$

as the virial for a mole. The factor $\frac{1}{2}$ appears because we have counted each pair of molecules twice. By partial integration we obtain

$$w = \frac{2\pi L^2}{v} 3 \int_\sigma^\infty U(r) r^2 \mathrm{d}r + \frac{2\pi L^2}{v} \sigma^3 U(\sigma) \, .$$

Introducing the abbreviations

$$w' = \frac{2\pi L^2}{v} \sigma^3 U(\sigma) \qquad \text{and} \qquad a = -2\pi L^2 \int_\sigma^\infty U(r) r^2 \mathrm{d}r \, ,$$

we then obtain

$$w = w' - \frac{3a}{v} \, .$$

c. Virial for the collision forces

The change in momentum of the two molecules contributes $2\sigma m w_N$ to the virial per collision, where $\boldsymbol{w}$ is the relative velocity of the colliding molecules. Since, for a given vector $\boldsymbol{w}$, $\pi\sigma^2 w_N \cdot n$ molecules collide per second with a given

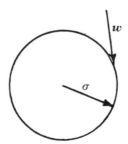

Figure 30.2

molecule, the virial is

$$2\sigma m w_N \, n\pi\sigma^2 w_N$$

per molecule. The time average of this is

$$2\pi n\sigma^3 m \overline{w_N^2} \, .$$

Since $m\overline{w_N^2} = 2kT$, the virial per mole is

$$w = \frac{L}{2}\, 4\pi\sigma^3 nkT = 2\pi\sigma^3 nRT \qquad (L/2 \text{ pairs of molecules}) \, .$$

Substituting $n = L/v$ into this relation, we have

$$w = 2\pi\sigma^3 \frac{L}{v}\, RT = \frac{3b}{v}\, RT \, ,$$

where $b = 2\pi\sigma^3 L/3$ equals half of the total volume of all interaction spheres in a molar volume.

d. Generalized barometer formula

Let the potential $U(r)$ be represented as

$$U(r) = \frac{L}{v}\, e^{-\frac{u(r)}{kT}} \, .$$

Then,

$$w = -\frac{L}{2}\int_0^\infty U(r)\, \frac{\mathrm{d}u(r)}{\mathrm{d}r}\, 4\pi r^3 \,\mathrm{d}r$$

$$= -\frac{2\pi L^2}{v}\int_0^\infty \frac{\mathrm{d}u(r)}{\mathrm{d}r}\, e^{-\frac{u(r)}{kT}}\, r^3 \,\mathrm{d}r \, .$$

Performing a partial integration, we obtain

$$w = \frac{2\pi L^2 kT}{v}\int_0^\infty r^3 d\left[e^{-\frac{u(r)}{kT}} - 1 \right]$$

$$= \frac{2\pi L^2 kT}{v}\left\{ \left[r^3\left(e^{-\frac{u(r)}{kT}} - 1 \right) \right]_{r=0}^{r=\infty} - 3\int_0^\infty r^2 \left[e^{-\frac{u(r)}{kT}} - 1 \right]\,\mathrm{d}r \right\} \, .$$

Defining

$$A(T) = 2\pi L^2 kT \int\limits_0^\infty r^2 \left[e^{-\frac{u(r)}{kT}} - 1 \right] dr \,,$$

we can write

$$w = -3 \frac{A(T)}{v} \,.$$

For the equation of state we thus obtain

$$3RT - 3pv - 3\frac{A(T)}{v} = 0 \,,$$

$$\left[p + \frac{A(T)}{v^2} \right] v = RT \,, \quad p = \frac{RT}{v} - \frac{A(T)}{v^2} \,.$$

The potential $U(r)$ has the form shown in Fig. 30.3, with

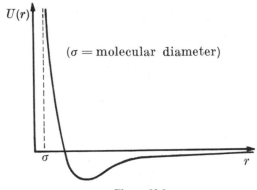

$(\sigma = \text{molecular diameter})$

Figure 30.3

the boundary conditions $U(r)|_{r=\sigma} = \infty$ and $U(r)|_{r\to\infty} = 0$. These conditions imply

$$\left(e^{-\frac{u(r)}{kT}} - 1 \right) \sim -\frac{u(r)}{kT} \quad \text{for } r \gg \sigma \,,$$

and

$$\left(e^{-\frac{u(r)}{kT}} - 1 \right) \sim -1 \quad \text{for } r \ll \sigma \,.$$

We have defined

$$A(T) = -2\pi L^2 kT \int_0^\sigma r^2 \, \mathrm{d}r - 2\pi L^2 \int_\sigma^\infty r^2 \, u(r) \, \mathrm{d}r \; ;$$

if we further define

$$a = -2\pi L^2 \int_\sigma^\infty r^2 \, u(r) \, \mathrm{d}r \quad \text{and} \quad b = \frac{2\pi\sigma^3}{3} L \,,$$

then we can write $A(T) = a - bRT$. Therefore, we have

$$p + \frac{a}{v^2} = \frac{RT}{v} + \frac{bRT}{v^2} = \frac{RT}{v}\left(1 + \frac{b}{v}\right),$$

which is an approximation to the van der Waals equation,

$$p + \frac{a}{v^2} = \frac{RT}{v-b} = \frac{RT}{v}\left(1 + \frac{b}{v} + \left(\frac{b}{v}\right)^2 + \dots\right).$$

Empirically one finds $a > 0$; that is, for $r \gg \sigma$, we have $U(r) < 0$, which implies attraction between the molecules. The radius of the interaction sphere σ can be estimated from these formulas and from the formulas for the free path.

Bibliography

Thermodynamics

O. SACKUR, *Lehrbuch der Thermochemie und Thermodynamik* (Berlin, 1928).

G. W. LEWIS and M. RANDALL, *Thermodynamics* (revised by K. S. Pitzer and L. Brewer) (McGraw-Hill, New York, 1961).

J. D. VAN DER WAALS, *Lehrbuch der Thermodynamik in ihrer Anwendung auf das Gleichgewicht von Systemen mit gasförmig-flüssigen Phasen* (Lecture notes edited by Rh. Kohnstamm in Amsterdam, 1908–1912).

W. H. SCHOTTKY, *Thermodynamik* (Berlin, 1929).

J. W. GIBBS, *The Collected Works of J. Willard Gibbs*, Vol. I: Thermodynamics (Yale University Press, New Haven, 1928).

M. PLANCK, *Vorlesungen über Thermodynamik* (Verlag von Veit & Comp., Leipzig, 1913).

P. S. EPSTEIN, *Textbook of Thermodynamics* (John Wiley & Sons, Inc., New York, 1937).

R. J. E. CLAUSIUS, *The Mechanical Theory of Heat* (Macmillan and Co., London, 1879).

Lectures on theoretical physics given by H. A. LORENTZ, HELM-HOLTZ, and KIRCHHOFF.

Kinetic Theory of Gases

O. E. MEYER, *Die kinetische Theorie der Gase* (Maruschke und Berendt, Breslau, 1877; second edition, 1899).

L. BOLTZMANN, *Vorlesungen über Gastheorie* (Verlag von Johann Ambrosius Barth, Leipzig, 1895).

A. KRÖNIG, *Grundzüge einer Theorie der Gase* (Berlin, 1856).

R. J. E. CLAUSIUS, various works.

J. C. MAXWELL, *Theory of Heat* (Longmans, Green, and Co., London, 1902).

J. JEANS, *An Introduction to the Kinetic Theory of Gases* (University Press, Cambridge, 1940).

E. H. KENNARD, *Kinetic Theory of Gases* (McGraw-Hill, New York, 1938).

H. A. LORENTZ, *Lectures on Theoretical Physics*, Vol. I: Kinetical Problems (Macmillan & Co., Ltd., London, 1927).

S. CHAPMAN and T. G. COWLING, *Mathematical Theory of Nonuniform Gases* (University Press, Cambridge, 1952).

See also works by MÜLLER-POUILLET and HERZFELD.

Appendix. Comments by the Editor

[A-1] (p. 2). What is meant here is thermodynamic equilibrium *in contact with another system at fixed temperature* (reservoir).

[A-2] (pp. 2, 5, 25, 29, 34). Actually, there are n relations (equations of state) which may also be written as $y_i = y_i(x_1, ..., x_n, t)$. Here in addition $t \equiv x_{n+1} = $ constant. An example with $n = 2$ is given on p. 29: Solving the equation of state of each subsystem for t yields $F(p, V, \bar{p}, \bar{V}) \equiv t(p, V) - \bar{t}(\bar{p}, \bar{V}) = 0$.

[A-3] (pp. 3, 21). Observation of the variation of temperature t by variable contacts with two reservoirs at different temperatures t_1 and t_2 tests the *monotony* of the scale between t_1 and t_2, but the *sign* of $t_1 - t_2$ is not determined. Indeed, this sign defines the direction of the heat flow according to the second law.

[A-4] (p. 11). Here Eq. [7.2] is meant to be taken at p_0.

[A-5] (pp. 12, 35, 37). In the following a quantity of substance of 1 mole or, as in Section 3, of 1 g should be considered.

[A-6] (pp. 21, 24, 33). The thermodynamic temperature scale is determined only up to a factor of arbitrary sign. The choice of the positive sign is a convention. Within this convention negative temperatures are uncommon although not inconceivable, as demonstrated by the work of Pound and Purcell, *Phys. Rev.* **81**, 279 (1951), on magnetic population inversion, and that of Ramsey quoted on p. 25.

[A-7] (p. 22). Actually, $Q_0 = 0$ by construction, since the coupling of the cycle to the reservoir at T_0 was only an artificial (and even unnecessary) device.

[A-8] (pp. 24, 34). On p. 23 it was proved that the entropy of a closed system cannot decrease. But here allusion is made to the arbitrariness of the sign of thermodynamic temperature (see [A-6]).

[A-9] (p. 30). This should read *quasi-static adiabatic*. For only in this case are the y_k in δW well-defined functions of $x_1, ..., x_n, t$. Note that the existence of at least one quasi-static path between any two points 1, 2 in state space is assumed here as well as on p. 23.

[A-10] (p. 45). From the free energy the stable isotherms can be determined without making use of these unstable states [see, e.g., K. HUANG, *Statistical Mechanics* (John Wiley & Sons, Inc., New York, 1963), Fig. 2.11].

[A-11] (pp. 61-72). This part from here to the end of Section 17 is an insertion into the second German edition of a manuscript written by Pauli in 1958 and which served as a basis for Pauli's paper in honor of J. Ackeret, *Z. angew. Math. Phys.* **9b**, 490 (1958). This paper is noteworthy for being Pauli's last published work [see "Bibliography Wolfgang Pauli" by C. P. ENZ in *Theoretical Physics in the Twentieth Century, a Memorial Volume to Wolfgang Pauli*, edited by M. Fierz and V. F. Weisskopf (Interscience Publishers, Inc., New York, 1960), and in *Collected Scientific Papers by Wolfgang Pauli*, edited by R. Kronig and V. F. Weisskopf (John Wiley & Sons, Inc., New York, 1964)]. The footnotes on pp. 64, 78 have been inserted at the same time.

[A-12] (p. 98). Actually, this result follows from Eq. [24.1] for a surface F chosen parallel and infinitesimally close to the wall W and for stationary conditions.

[A-13] (pp. 119, 120). Here $d\Omega$ and $d\Omega'$ are the elements of solid angle through which the surface elements $d\sigma'$ and $d\sigma$ are seen from P and Q, respectively, i.e., $(\overline{PQ})^2 d\Omega = \cos\vartheta' d\sigma'$, $(\overline{PQ})^2 d\Omega' = \cos\vartheta d\sigma$. [This is the same situation as in Section 4 (photometry) of the volume *Optics and the Theory of Electrons* of this series.] The radiation from Q toward P is by definition $A_Q \cos\vartheta' d\sigma' d\Omega'$. Integration of this quantity on p. 120 is over the full solid angle Ω, and not just over the upward half as on p. 119. Therefore the term $A_P \cos\vartheta \sin\vartheta\, d\vartheta\, d\varphi\, d\sigma$ integrates to zero.

Index